AF443598

What SOLAR ENERGY can do for you

What
SOLAR ENERGY
can do for you

Ronald C. Denney

FREDERICK MULLER
LONDON

First published in Great Britain 1980 by
Frederick Muller Limited, London NW2 6LE

British Library Cataloguing in Publication Data

Denney, Ronald C.
What solar energy can do for you
1. Solar energy
I. Title
621.47   TJ810
ISBN 0 584 10436 7

Set by Ⴠ Tek-Art, Croydon, Surrey.
Printed in Great Britain by The Garden City Press Ltd

To my parents
for
life, faith, hope and love

CONTENTS

List of Figures		ix
List of Plates		xi
Preface		xv
Introduction		xvii
Chapter 1	Why Solar Energy	21
Chapter 2	The Sun As An Energy Source	31
Chapter 3	Using Solar Energy	41
Chapter 4	Large Scale Use of Solar Energy	75
Chapter 5	Electricity from Solar Energy	86
Chapter 6	The Future Role of Solar Energy	95
Appendix		104

LIST OF FIGURES

1. Solar Radiation Energy Distribution Curves — 23

2. World Insolation Distribution — 34

3. Diffuse and Direct Solar Radiation Contribution at Kew, London — 35

4. Effect of Collector Angle on Amount of Incident Solar Radiation — 36

5. Focusing Effect of Parabolic Concentrators — 39

6. The 'Greenhouse' Effect used to assist House Heating — 42

7. Air Circulation Diagram for the Passive Solar System at Acorn Close, Higher Bebington — 45

8. Passive Heating from a Window Frame Unit — 46

9. Main Design Features of a Flat Plate Collector — 47

10. Tube Layouts for Flat Plate Collectors — 50

11. Roof Arrangements for Flat Plate Collectors — 50

12. Cross-Sections of Tube and Absorber Designs — 51

13. Circulation for Thermosyphoning Domestic Solar Water Heating System — 53

14. Circulation for Pumped Solar Domestic Water Heating System — 55

15. Circulation System for Solar Heating System at the Students' Flats, Llys Tal-Y-Bont, Cardiff, Wales — 57

16. Water Circulation and Heat Storage System in the Milton Keynes Solar House — 61

17. 'Bread Box' Type of Solar Water Heater — 63

18. Solar Collector Design for Swimming Pools — 65

19. Principle of Solar Distillation — 69

20. Cold Frame Type Solar Still — 69

21. Simple Sand Pit Solar Still — 71

22. Vertical Solar Still — 71

23. Solar Oven — 72

24. Solar Drier — 73

25. Parabolic Concentrator for Steam Generation — 78

26. Principle of Stirling's Heat Engine — 80

27. Heat Pump Circuit — 82

28. Combined Heat Pump-Solar Panel Circuit — 84

29. P/N Thermocouple — 87

30. Tandem Cell — 90

31. Solar Space Station Concept — 99

LIST OF PLATES

I   Radiometer   29

II   St. George's Middle School, Wallasey   43

III   Acorn Close, Higher Bebington   44

IV   Solar Roof Panels at   52
      (a) Fairseat, Kent   (c) Basing, Hampshire
      (b) Sevenoaks, Kent   (d) Southwark, London

V   Solar Panels on Staff Cottages at the Centre for   54
      Alternative Technology, Machynlleth, Wales

VI   Wates Houses at Crofters Mead, Forestdale, Croydon   56

VII Students' Flats at Llys Tal-Y-Bont, Cardiff, Wales   56

VIII Solar Panels on the Roof of the Mountain Hotel,   58
      Brecon, Wales

IX   Milton Keynes Solar House   60

X   The Duke of Kent Visiting the Solarsense Stand   62
      at the British Energy Exhibition in Peking in June 1979

XI   Solar Heating Unit for the Swimming Pool at   64
      Ye Olde Felbridge Hotel, East Grinstead, Sussex

XII Solar Panels used for Fish Farming at the Centre   67
      for Alternative Technology, Machynlleth, Wales

XIII Plastic Still used by the Armed Forces for   70
      Emergency Purposes

XIV Solar Furnace at Odeillo, France   77

XV   Solar Cell Array   89

XVI Solar Cells in use   91

Plate VIII is reproduced by permission of the proprietors of the Mountain Hotel; Plate X is reproduced by permission of Solarsense Ltd. Plates XIII and XIV have been kindly provided by the R.A.F. and the Keystone Press Agency Ltd. respectively, and Plates XV and XVI by Lucas Services Overseas Ltd. All other photographs are the copyright of Dr. Ronald C. Denney.

Our doubts are traitors, making us lose the
good we oft might win, by fearing to attempt.

William Shakespeare
Measure for Measure I iv 77

PREFACE

During the preparation of this book I have been very conscious of the considerable debt I owe a large number of people who have provided data, advised me and spent their time showing me around solar installations. I have been particularly impressed by their enthusiasm and general willingness to pass on their experiences. I hope that this book does them justice.

I must, however, thank my very good friend Alistair Craig, B.Sc. (Eng.), C.Eng. for many helpful discussions, the provision of article reprints and the loan of equipment. Special mention must be made of Dr. Max Johansson of the Solar Energy Unit, University College, Cardiff for helpful discussion and permission to use data published in *Helios*. I am grateful to Alan Horton of the Polytechnic of Central London for spending so much time showing me over the Milton Keynes house and going over the energy data with me. I must also include Adrian Sturgeon, B.Sc.(Eng.), A.C.G.I., M.I.C.E., for so willingly supplying details of the installation and running costs of his solar panels and Hugh Pasteur for showing me his long-serving heat pump.

I have had nothing but help from all sides of British industry when I have telephoned or called for information and advice. Companies such as Pilkington Brothers Ltd., Lucas Service Overseas Ltd., Wednesbury Tube Company Ltd., Solarsense Ltd. have all given generously of their time and brochures. They have willingly given permission to reproduce diagrams, tables and photographs which have added to the quality of this book.

Once again I must acknowledge the considerable help I have received from the librarians at Thames Polytechnic in obtaining the most obscure journals for me to study. Their tolerance, understanding and willingness to help has made my job much easier.

Finally I must thank my wife and children, who have once again accepted my absences from home and my late-night writing as part of a normal life. They have also put up with the moods of the frustrated author whose brainchild takes precedence over most other responsibilities. I only hope that all their sacrifices will be rewarded by this book having a wide appeal for the general reader for whom it is intended.

Ronald C. Denney

INTRODUCTION

As the international price for crude oil continues to rise due to the oil-producing nations' attempt to increase their incomes from a wasting asset, so the search for alternative sources of energy has been intensified. Every new idea is initially heralded as the potential energy saviour of the world until the euphoria eventually disappears under the problems that inevitably arise when reality is reached. It is, however, quite clear that there is no single solution to the energy problem and all possible approaches must be evaluated and considered. Developing and expanding economies require ever-increasing supplies of energy as the standard of living is directly related to the energy consumption of a nation. Even if the developed countries cut back on their enormous demands, any temporary surplus will very soon be taken up by the underdeveloped nations as their economies advance, and the need for alternative energy sources will remain.

The real problem is not whether or not oil reserves will last for thirty, forty or fifty years, or even beyond that, but what the world is doing *now* to ensure that there are adequate supplies of energy at all times. In the case of oil it is not simply a question of replacing it as an energy source but also of finding alternatives for the oil-based animal feedstuffs, the raw materials for the chemical industry, the liquid fuel for petrol engines, and for the energy source for electricity generation. When considered in this way it should be quite clear that it is unlikely that any single *new* energy source is likely to solve all these requirements at one and the same time.

Having said that, it *is* theoretically possible that solar energy could solve all these problems in the sense that photosynthesis *will* provide the animal feedstuffs, that fermentation *could* provide the chemical raw materials and the engine fuel, and that it *could* be used for the production of electricity. But theory does not always work out in practice and those journalists and broadcasters who have implied that all problems will be solved simply by investing vast sums of money into solar energy research are guilty of both misrepresentation and of the failure to acknowledge that solar energy research has actually been proceeding successfully in various countries for many years. There is no doubt that solar energy is capable of providing a much greater proportion of the world's energy needs than has hitherto been the case, but the transition to that state will take a long time and at the same time will require a complete alteration in people's attitudes and mental states. If solar power is to become generally accepted as an alternative energy source it must not only be economically competitive with other energy sources but must also be relatively easily installed, whilst the equipment must possess a reasonably long life with low maintenance costs and servicing requirements.

There is, too, a need for a greater appreciation by the public that solar energy can benefit everyone and that it is not necessary to live within a few miles of the equator to be able to harness the sun's power. There is no doubt that in Great Britain many people were totally amazed in July 1979 when the Post Office announced that it had installed the first solar-powered telephone box on a remote wind- and rain-swept island near Skye in Scotland. If solar energy can work under those conditions, it should work almost anywhere else in Britain and in other countries with so-called temperate climates.

As fossil fuel prices rise so the alternative energy sources become more economically viable, the only problem being that to get the new energy source functioning in the first place requires the use of materials and products which are initially produced using the high-cost fossil fuels. As a result the costs of obtaining the energy from alternative sources also increase partially in line with the established energy sources, and the breaking of the vicious circle is far from easy.

There are many reasons for writing a new book on solar energy. As a technical author I have always been concerned that scientists and technologists tend to live in worlds of their own, using languages remote from people around them. And all too often when communication and comprehension are difficult the scientists place the blame on other people and not on their own inability to express themselves in such a manner that others can understand. Because of this I feel that part of my job is to try to explain scientific developments in a style that the non-scientist can comprehend.

An additional reason for writing the book is to increase the public awareness of the subject, of its potential benefits and of its suitability for domestic application. I hope that what I have to say will lead many people to look into the possibilities of solar energy for their own homes and into trying out some simple solar energy devices just in order to study how effective they can be even in a climate not renowned for its sunshine. I have tried to be objective in my assessment of the subject and to draw attention to its limitations as well as its benefits.

It is only fair to point out that most of the fundamental problems associated with the application of solar energy have already been solved. What is taking place now in the research centres and universities is a refinement and improvement on techniques and methods which were originally developed some years ago. New materials which have been manufactured in recent years have meant that it is now possible to obtain greater efficiencies from solar energy conversion, and the scope for its application is rapidly expanding. Whilst it is unlikely that it will ever totally displace other sources of energy it has a definite role to play in helping to provide the needs of an energy-hungry world.

1

WHY SOLAR ENERGY

It is only a few thousand years since the energy needs of primitive man were limited to obtaining enough wood or animal dung to provide him with fuel for his fire to keep warm at night. Certainly in those days the energy problem and the pollution problems were far removed from those now being experienced in the developed world. At the present time, if the industrialised countries fail to find long-term solutions to the problems that confront them, this will in turn hold back the development and elevation of the living standards in the poorer and underdeveloped parts of the world. Those people who criticise the U.S.A. for the fact that 6 per cent of the world's population use more than 30 per cent of its energy always imply that if the U.S.A. reduced its energy load it would solve all the world's problems overnight. It would, of course, do nothing of the sort. All energy conservation programmes can achieve is an extension to the potential lifetimes of fossil fuels during which other alternative energy sources may be better developed.

The problem being experienced by the developed world is due to the fact that in an advanced technological state the average person is using about one hundred times the amount of energy that primitive man used and about three times the amount that would have been used even in an industrialised society a hundred years ago. This means that energy demands during the past fifty years have placed an enormous pressure on all resources and any long-term reduction in available energy supplies, even allowing for conservation programmes, will ultimately mean a reduction in living standards for the developed world and reduced expectations for poorly developed nations.

Over the last one hundred years there has been a progressive transition away from the almost exclusive use of wood as a fuel, to coal in the late nineteenth and early twentieth century and then to oil during the last fifty years. At the same time, where possible, countries have developed hydroelectric power, the use of natural gas (another fossil fuel), and even more recently nuclear power. During this time the use of solar power, wind power, tidal flow systems and terrestrial heating has been virtually negligible in the energy equation. It is because the supplies of fossil fuels are finite with lifetimes that can be fairly accurately calculated that the time for alternative energy sources has now arrived. Although in some cases it may be many years yet before the technological break-throughs are made, it is important that the study, research and investigations are carried out now, well before the oil and gas run out, rather than left until it is too late. At the same time it is equally important that proven alternative energy sources be utilised as fully as possible in order to maximise the lifetime of the fossil fuels.

To put the matter more in perspective, it has been estimated that the world's known and potential reserves of crude oil will be exhausted within about 35-40 years, and that extraction will progressively decline during the first fifteen years of the next century. Natural gas reserves will last for only fifty years and coal for about three hundred years. From this it can be seen that what-ever oil crises may now be experienced are nothing compared with those likely to arise during the lifetimes of our children and grand-children. It should also be pointed out that there is little likelihood of hydroelectric power taking over a much more substantial role, as it has already been well developed where possible and currently provides about 5 per cent of the world's energy needs.

There are very few nations in the world that can be considered to be totally self-reliant for energy. Within the European Economic Community nearly 60 per cent of the energy used presently comes from outside the E.E.C., and after various conservation measures the figure will still be 50 per cent in 1985. By that time only Great Britain and Holland within the E.E.C. will be self-reliant for energy (as a result of North Sea oil and gas), and then only for a few years.

In the same way the energy balance in the U.S.A. is nothing short of disastrous. Despite substantial oil deposits, large coal reserves, hydroelectric schemes and ample solar radiation, the U.S.A. now imports 40 per cent of its oil (that is about 20 per cent of its total energy requirement). Only a total readjustment of attitude towards the use and misuse of energy can lead to the U.S.A. becoming self-sufficient in energy once again.

It is for these reasons that man's mind and ingenuity have been turned to the possibility of harnessing a larger proportion of the limitless supply of free energy coming from the sun. The actual amount of radiation reaching the earth's surface, at 30,000 times the total daily need of energy, is so large as to be almost beyond comprehension. The upper atmosphere receives about 17.3×10^{13} kW of solar radiation covering a range of wavelengths from the ultra-violet through to the infra-red (this distribution is shown as the upper curve in Figure 1). Some of the radiation is reflected from the atmosphere back into space and another portion is absorbed by the atmosphere itself so that the amount of useful energy falling on the earth's surface may be less than half of that reaching on the outer atmosphere (the difference in the energy distribution is shown by the lower curve in Figure 1) and will in any case vary with the amount of cloud. However, even allowing for this the amount of energy is far in excess of what would normally be required, if only it could all be harnessed. Even when skies are overcast with cloud, quite a lot of diffuse sunlight still reaches the earth and can be used by solar collectors. In many ways the sun's power has been very much taken for granted. We see it in the sky, we feel its warmth on our faces, we see it ripen our crops. Because it is there all the time we have tended to ignore its real potential in our lives.

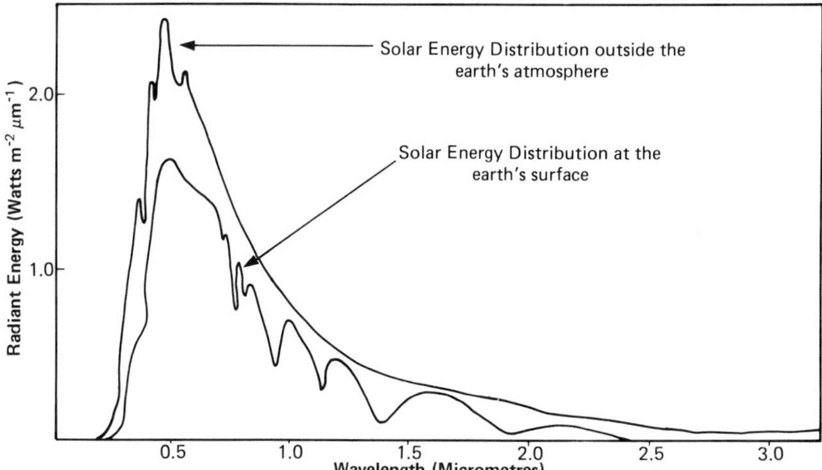

Figure 1.   Solar Radiation Energy Distribution Curves.

The sun has fascinated people ever since man first appeared on this planet, although sun worship on the beaches takes a different form than it did with the ancient Greeks and Romans. The sun has been given many names as a god by various civilisations. The ancient

Egyptians had at least five different names depending upon the state of the sun, the most well known being Ra, associated with the sun at its zenith. To the Greeks it was Helios, from which we obtain our word helioscope, although it was Apollo that the Greeks considered to be the God of solar energy and he was also accepted as a god by the Romans. In Indian mythology the god responsible for causing the sun to cast its rays on the earth is Savitar, but the deity of the sun itself is Surya, to whom twelve different names are ascribed depending upon the practical role with which the sun is associated.

This worship of the sun did not, of course, prevent man from making use of the sun for his own benefit and simple applications of solar radiation have existed for thousands of years. The drying of crops, baking of bricks in the sun and the drying of washing are all examples of the use of solar radiation which are still used throughout the world today. The growth of crops to produce food and wood for fuel are both conversions of solar radiation into forms that can be used by man, and what are known as 'biomass' experiments are now being carried out to try and make even these processes more productive, as the process of photosynthesis by which plants use solar energy is normally only about 1 per cent efficient.

Apart from the traditional uses of solar radiation, the ancient philosophers and their contemporaries did devise other applications for solar energy. Socrates is reported to have designed a house with a large roof overhang in order to cut out the direct rays of the high summer sun but to maximise the use of the lower winter sun. Similarly the Greek historian Xenophon advised that houses in the northern hemisphere should be built facing the south with the roof pitched down south to north in order to catch the winter sun. Concepts such as these are now being incorporated into the various experimental solar houses being constructed throughout the world .

Other early uses for solar radiation include the distillation of alcohol from fermented fruits, a procedure believed to have originated with the Arabs. It was Euclid who first described the focusing effect of spherical mirrors; an observation which was put to a highly destructive use by Archimedes, who used concave mirrors of large curvature to focus the sun's rays on the sails of the ships of the Roman general Marcus Claudius Marcellus at Syracuse, setting fire to them and breaking the siege. Although this claim has been criticised as a myth it was shown to be feasible in 1747 by a Frenchman, M. Buffon, who used more than 150 mirrors at a focal point of over 250 ft. (75 m) to set fire to wood and charcoal. However, from a really practical point of view most of the important advances in the use of solar radiation have only occurred during the last three centuries, and particularly since the middle of the nineteenth century. Between the early seventeenth and late eighteenth centuries, various attempts were made to invent solar-powered pumps and studies were also made on the concentrating powers of curved surfaces. In 1774 Joseph Priestley focused the sun's rays through a large lens onto mercuric oxide and discovered that it released a gas. Although he was unable to explain fully the chemical processes involved in the reaction, it was this study that led to his discovery of oxygen. At about the same time the French scientist Lavoisier developed a solar furnace with a lens nearly two metres in diameter.

With the coming of the industrial revolution and the general growth of industry, real progress in the application of the sun's energy took place in the mid-nineteenth century. The most important advances were aimed at developing solar engines, and the first book on solar energy was written by a French Professor, August Mouchot, and published in Paris in 1869. It was the availability first of coal and later of oil as cheap sources of fossil fuels that led to the decline in the interest of solar power for low-powered engines, a situation which is only being slowly reversed at this time. However, solar engines were still being developed in the U.S.A. during the early part of the twentieth century and in 1913 a 50 h.p. pump was successfully built for pumping irrigation water in Egypt.

Another major development that took place over one hundred years ago was the use of solar energy for distillation to produce drinking water from salt water in Chile. The solar unit for this process covered about 50,000 ft^2 (4,700 m^2), roughly the size of a football pitch, and consisted of sloping glass roofs over shallow pans of water. The system was capable of producing 6,000 gallons (27,000 litres) of water each day and operated for forty years until the mining in that isolated area was discontinued. The method used in this operation remains the basis of solar distillation to this day.

Other early developments have included various forms of solar cookers and solar furnaces, solar-operated refrigerators and solar-heat reservoirs. It is not surprising that with all these developments the first solar power company was founded in California in 1883. However the development of oil and all its products prevented any major extension of research into the uses of solar power and for many years only a very few persistent dogged individuals worked in this field.

For many years very little work on solar energy was carried out in Great Britain and people like Dr Harold Heywood, Principal of

Woolwich (now Thames) Polytechnic, who built and installed his own solar collectors at his home in Sidcup in the 1950s were pioneers in a badly neglected area of work. It is, however, worth bearing in mind that at this time the first commercial solar water heaters were already available in Israel, and as a result that country became one of the leading users of solar energy for domestic hot water. The fact that Israel enjoys an average of 3,500 hours of sunshine each year has been a great incentive to its solar energy trade. Similarly, simple types of solar domestic water heaters using glass-topped boxes lined with black plastic were developed in Japan in 1948. By the early 1970s they had been fitted to 30 per cent of the homes in that country. With a normal cold water inlet they are capable of providing hot water at 55°C (131°F) in summer and above 30°C (86°F) in winter.

Internationally the flat period for academic research on solar energy existed for about twenty years until just before World War II. During this time applications of solar power were limited to some work on solar-powered steam engines and partial heating of family residences in the cooler part of the year in Florida. The change, mainly in academic attitudes to solar research, can be virtually dated from the decision by the Massachusetts Institute of Technology (MIT) to build its first 'Solar House' in 1939. Since 1940 the number of solar houses constructed in the U.S.A. has exceeded one hundred, whilst more than twenty had been built in Great Britain by 1976. In the past five years this research has been enormously extended as France, Germany, Japan and Australia have introduced major research programmes in this technical field, and the number of solar houses, both experimental and for normal habitation throughout the world, totals several hundreds.

Following their first Solar House, which used a 30°, south-sloping roof with 366 ft² (34 m²) of solar collector, MIT improved its design of solar houses and by 1959 had developed a house in which 50 per cent of the total heating was provided by solar energy over a two-year period. By this time the first solar-heated office building had also been built, at Albuquerque, New Mexico, with 63 per cent of its heating being provided by solar radiation, from 750 ft² of flat-plate solar collectors providing heat to a 6,000 gallon water storage tank. It was only shortly prior to this date, in 1956, that the first solar house was constructed privately in Great Britain. This particular house, which was designed by the owner at Rickmansworth, Herts., had the south side of the building almost completely faced with double-glazed solar panels.

The early experimental solar houses did not attain the perfect

26

state of being entirely autonomous throughout the whole year and required supplementary heating during the coldest weather. It was not until 1975 that the Copper Development Association in Tucson, Arizona, constructed a house that not only achieved 100 per cent heating from solar radiation but also used it to power 75 per cent of the air conditioning and to charge a low-powered electric vehicle. The electricity was generated by the use of silicon-based solar cells. Although it would be wrong to assume that this type of energy independence can be automatically mass-produced for the whole of the developed and under-developed world, what this construction did show was that all the technology and materials necessary already exist for this purpose, and that further technological advances should make such systems much more widely available.

More recent applications have included a factory and office complex, built in 1978, in Zurich, Switzerland. This is one of the largest solar buildings in the world, providing not only all the hot water but also some of the space heating for the four-storey building. A point well worth bearing in mind is that the solar installation represented less than 2 per cent of the total building costs, and actual energy requirements have been reduced by good insulation and construction with low thermal conductivity materials.

The 1973 oil crisis, in which the main oil exporters quadrupled the price of oil almost overnight, elevated the study of solar energy from being simply an academic exercise or a rich man's toy to the role of a potentially serious contributor to the energy supply. At the same time, it is only fair to point out that people with vision had realised this potential much earlier. In fact, the first major international symposium on wind and solar energy had been held in New Delhi nearly twenty years earlier, and that in turn led to the growth of the International Solar Energy Society (ISES), which. now has sections in most of the major countries in the world.

The current position is that very large sums of money are being provided for further research into solar energy utilisation and some governments are looking at the possibility of providing state grants to encourage people to install solar panels for domestic water heating and to supplement central heating systems. As part of the United States energy programme, President Carter had 32 solar panels installed in the roof of the White House at a price of about $28,000 (£14,000) in April 1979, with the intention of providing about 75 per cent of the hot water used in the West Wing. This was the first step in the attempt (considered by many to be highly optimistic) to make the U.S.A. obtain 20 per cent of its energy needs from the sun by the end of the twentieth century.

All nine member states within the European Economic Community have embarked upon major energy research and development programmes in which solar energy figures prominently. It is proposed that by 1985 reliance upon imported energy sources will be reduced from the 1976 level of 60 per cent to about 50 per cent. Whilst it is not expected that solar energy will be solely responsible for this, there is no doubt that it is hoped that it will play a major role in the domestic situation. Of major interest in the E.E.C. are projects concerned with the storage of solar energy and its conversion to electric power. The funds allocated for research on solar energy in the E.E.C.'s second energy programme for 1979-1983 now stand at £29 million. Within the United Kingdom more than £6 million will be spent between 1977 and 1981 on research into domestic water and space heating, biomass conversion and photo-voltaic electricity production.

With the large amounts of money now available for solar energy research, more solar houses and new designs of collector panels are being devised. There is no doubt that this work will have an enormous effect on the nature of buildings designed in the near future.

One of the largest state financed programmes of research is being carried out in Brazil. There about £150 million per annum is being spent on biomass studies, using solar energy to grow sugar cane, sorghum and cassava that can be used to produce alcohol as a fuel to replace up to 20 per cent of petrol in motor vehicles. And even the oil-rich Arab states are using some of their oil revenues for solar energy programmes aimed at producing better water desalination plants and air-conditioning systems. Within Great Britain solar power has until recently received a very low priority, probably because of the scepticism that greets anyone who suggests that the British weather can be used to anyone's advantage. At the time of writing 50 per cent of all solar energy research and development is being carried out in the U.S.A., and most other countries are still a long way behind in effort but not in ideas and originality. The great danger is always that the potential for solar energy will be over-stated. In real practical terms it is possible that solar power can provide between 5 and 10 per cent of the energy needs of the developed countries, and only a major technological breakthrough, for instance in electricity generation, will lead to this point being readily improved upon. Within the poorer countries, many of which are also the hottest, there is the additional possibility that solar power can provide the low-powered engines needed for irrigation and the water purification systems that will help to reduce disease. Within the following chapters the various approaches to these applications are discussed in order that the reader can evaluate them for himself.

For the ordinary person who has not seen a solar house or even noticed the occasional roof fitted with flat-plate solar collectors, it is necessary to demonstrate that the sun's energy can be collected, measured and put to useful work.

One of the simplest devices used to demonstrate the power of solar radiation is the radiometer invented by Sir William Crookes in the nineteenth century. It consists of a partially evacuated glass bulb, about the size of an electric light bulb, with a vertical pin on which are pivoted four small square vanes at right angles to each other (Plate I). Each vane is white on one side and black on the other. A number of explanations for the motion have been given, the most probable reason is the following: when sunlight, or any other radiation, is directed onto the glass bulb, the black sides of the vanes become hotter than the white sides. This is because black surfaces absorb solar radiation whilst white surfaces tend to reflect it. As a result of this, any gas molecules in the bulb which strike the black surfaces acquire a substantial motion in one direction and impart a corresponding opposite force to the surface. The continual

Plate I. Radiometer

29

impact and rebound effect of the molecules causes the vanes to rotate, the speed of rotation corresponding to the amount of energy falling upon the glass bulb. In direct sunlight the speed of rotation can exceed 2,500 r.p.m.

Another very simple demonstration of the sun's power is to take a 50-100 feet (15-30 m) length of black hosepipe and fit it to the cold tap on a hot day. Then stretch the length of the pipe out in the sunshine. If the cold tap is turned on at a moderate rate it will be found that the water issuing from the end of the hosepipe is hot due to the absorption of solar radiation during its journey along the pipe.

The most common experience most people have with solar radiation is the way in which a closed car becomes heated inside even on days in which the sky is overcast. This phenomenon is striking proof of the 'greenhouse' effect. The actual result with a car can be easily studied by using two thermometers, one placed outside the car to record the external temperature, and the other placed inside the car in a position from which it can be read. On a hot day it will be found that the temperature inside the car will rise to about 30°C (54°F) above the external temperature within an hour, and will stay above the external temperature for up to four hours after the sun has become covered by cloud.

All small boys are familiar with the use of a convex (converging) lens to concentrate the sun's rays either to burn someone's arm or to set fire to a piece of paper. It is this particular action of curved pieces of glass that has been responsible for the destruction of many acres of heath and woodland over the years, because of the sun shining through broken bottles or lost spectacles. The value of the use of curved lenses and mirrors to concentrate the amount of solar energy on one point has not been lost by the developers of solar energy equipment, although in practical terms it has been found that parabolic reflectors are normally superior to circular shapes.

Although these devices and experiments may appear quite simple, they actually form the basis of much of the solar energy equipment now on sale. A flat-plate solar collector is a glorified form of the hose-pipe, a solar furnace is an overgrown magnifying glass and a solar still is a wet form of greenhouse. Modern solar energy applications are based upon very old knowledge that can still be applied usefully and will continue to grow in value in the future.

2

THE SUN AS AN ENERGY SOURCE

Statements that claim that the Earth receives as much solar energy in ten days as is available from the total oil, coal and gas reserves in the world are pointless and meaningless. Although the amount of energy available from the sun is vast, not all the radiation reaching the earth's outer atmosphere is available for use by man. It is worth pointing out very early on that the usable amount of solar radiation depends upon the quantity of radiation that actually reaches the earth's surface, the efficiency of conversion by the solar collector and the area of the planet that is covered by solar collectors. Another economic feature that tends to be ignored is that there is no point in spending vast sums of money on expensive solar collectors if the cost in energy terms of producing the materials in the collectors is greater than the amount of energy that will be produced in the lifetime of the device.

The energy we receive is actually only a very small proportion of the total radiation produced as a result of the high temperature nuclear fusion reaction taking place at the sun's surface. It is believed that the high temperatures and the energy released arise due to the fusion of hydrogen nuclei to give helium nuclei with a small amount of mass being simultaneously converted to energy. This is the process that takes place in a hydrogen bomb, and it will be appreciated that the destruction of even a very small amount of matter releases a huge amount of energy. As the sun is losing mass at the rate of about 4×10^9 kg (4×10^6 ton) every second, the quantity of energy released in all directions is enormous. The nuclear fusion reactions result in temperatures at the sun's centre of the order of millions of degrees centigrade, whilst at the sun's surface more moderate temperatures of $5,500$-$5,700°C$ ($9,900$-$10,300°F$) are attained. The sun's radiation covers a very wide range of wavelengths characteristic of what is known as

31

'black body' radiation — that is, it is like the radiation obtained from heating a perfect black object at more than 5,500°C. The distribution of this radiation is shown in Figure 1 (page 23) but in practical terms the radiation reaching the earth's surface differs from this, due to the fact that some of the radiation is reflected back into outer space and some of it is absorbed by the carbon dioxide and water vapour in the atmosphere. As a result the radiation available for solar collectors is the lower, more ragged curve shown in Figure 1. The higher the sun is in the sky, the more intense the radiation that actually reaches the measuring point, and the lower it is in the sky the greater the journey through the earth's atmosphere and the smaller the intensity at the earth's surface. In any case, passage through the earth's atmosphere eliminates most of the ultra-violet radiation from the sun's rays as well as some of the infra-red radiation. However, in addition to the direct radiation at any point there is always some contribution due to diffuse light arising from radiation that has been scattered on its passage through the atmosphere. This varies greatly, but can be a significant contributor to the overall quantity of energy received by a solar collector, sometimes being as much as a half of the total. It is of particular value in countries such as Britain, where the skies are frequently overcast.

Unfortunately a great deal of confusion has been created by the use of different units for the amount of radiation reaching any spot on the globe. As a result it is not always easy to compare one country with another or to assess variations in solar energy throughout the year. Very simply the relationship between the units used is as shown below:

The unit of power is the watt, which is a measure of the amount of energy (joules) per second.

$$1W = 1J\ s^{-1} = 0.239 \text{ calories per second} = 9.48 \times 10^{-4} \text{ British Thermal Units per second.}$$

Solar radiation is sometimes expressed in langleys such that

$$1 \text{ langley} = 1 \text{ cal cm}^{-2} = 4.18\ J\ cm^{-2} = 3.68\ BTU\ ft^{-2}$$

However for most purposes it is sufficient to use various multiples of watts and joules and the other units need not be employed.

The outer atmosphere receives about 1.35 kWm^{-2}, i.e. 1.35 $kJm^{-2}s^{-1}$, and this is known as the 'solar constant'. Due to absorption in the atmosphere the actual maximum radiation reaching the Earth's surface is more of the order of 0.7 $kW\ m^{-2}$, and may be less than that due to cloud cover. This means that on a hot clear day with twelve hours of continuous sunshine the maximum insolation would be 8.4 kWh m^{-2}. For most parts of the world the actual maximum levels attained are less than this, varying enormously throughout the year. In winter months the highest amount of insolation may be only 10 per cent of that received in summer months.

From a domestic heating point of view this is round the wrong way: at the time when solar heating is needed most there is the least supply of it available, correspondingly there tends to be a surplus at the time of the year when it is least required. For this reason it is normally necessary to supplement the solar heating with other energy sources during winter months.

In the U.S.A. the annual number of hours of sunshine varies from about 4,000 hours per annum in parts of Arizona down to 2,200 in New England and Oregon. In Great Britain the corresponding figures are between 1,600 and 1,200 hours per annum, the higher values being recorded in south and south-west England. As would be expected, this is far less than is obtained in countries like Kenya (3,000 hours per annum), Portugal (2,800 hours per annum) and Brazil (2,300 hours per annum). Despite the apparently low figures for Great Britain, many people have managed to make solar energy work for them in providing a portion of their domestic heating, heating swimming pools or employing the 'greenhouse' effect.

The variation of solar radiation throughout the world and at different times of the year may be seen in Figure 2. One point of particular note is that countries at latitudes between 40° and 60° still receive about 50 per cent the amount of solar radiation that falls on countries closer to the equator.

The annual variation in solar radiation for south-east England in Table 1 is fairly typical of what can be expected in most countries away from the equator, although actual values will differ.

January	0.5 kWh m^{-2}	July	4.4 kWh m^{-2}
February	1.1 kWh m^{-2}	August	3.7 kWh m^{-2}
March	2.0 kWh m^{-2}	September	2.7 kWh m^{-2}
April	3.3 kWh m^{-2}	October	1.5 kWh m^{-2}
May	4.3 kWh m^{-2}	November	0.7 kWh m^{-2}
June	4.9 kWh m^{-2}	December	0.4 kWh m^{-2}

Table 1.   Mean Daily Insolation for South-East England

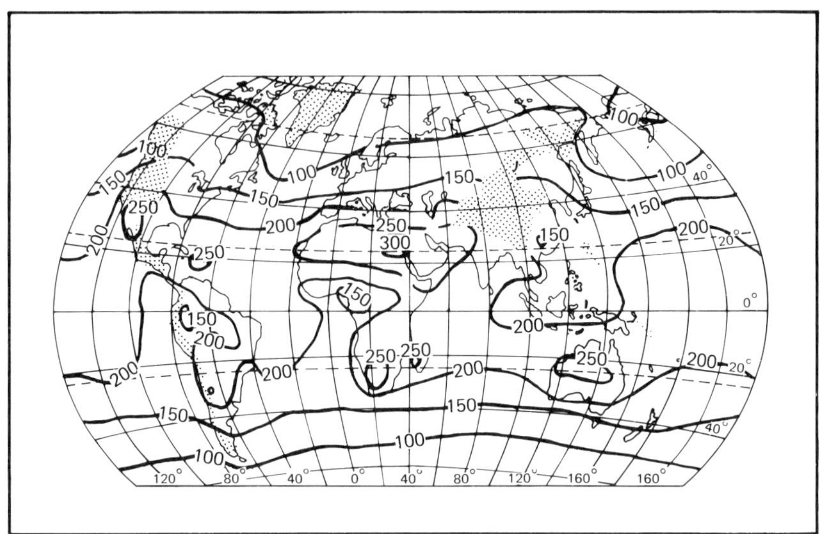

Figure 2.   Annual mean global irradiance on a horizontal plane at the surface of the earth (W/m² averaged over 24 hours).

Published with the permission of the UK section of the International Solar Energy Society from *Solar Energy, a UK Assessment,* May 1976.

The large amount of solar radiation may be used in a variety of ways, but in terms of domestic usage there are three main applications. These are for:

(a)   domestic hot water; for baths, showers, washing clothes and washing crockery.

(b)   space heating; either for ducted air systems or for heated water central heating circuits.

(c)   heating swimming pools.

Calculations show that in houses in Great Britain 64 per cent of domestic energy consumption is used for space heating and 22 per cent for water heating, the remaining energy requirements being for cooking, lighting and electrical appliances. In one year the 86 per cent for heating requires something of the order of 70 GJ (i.e. 70 x 10^9 J) and various studies have shown that the potential contribution of solar radiation in Great Britain is such that a saving of about 15 per cent could be made on the figures for the national primary energy consumption. Similar savings are possible in other countries.

Such projections must, of course, be related to the amount of available solar radiation. This is usually measured relative to a horizontal surface (although to maximise the solar collection

34

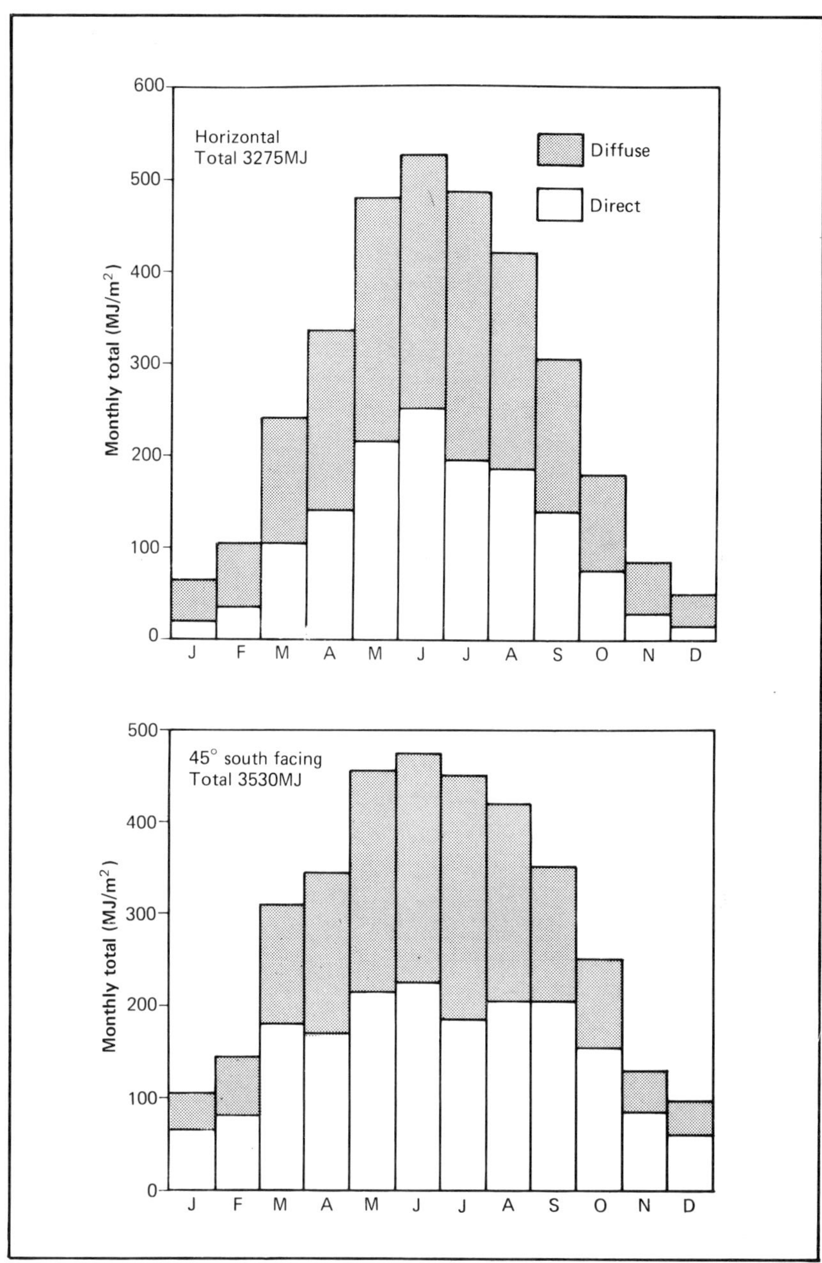

Figure 3.   Diffuse and Direct Solar Radiation Contribution at Kew, London

Reproduced by permission of The Building Research Station, Garston, WD2 7JR, from Building Research Establishment Current Paper, CP64/76.

surfaces are normally set at an appropriate angle to the sun's rays).
In Great Britain an average of 3.0 GJ falls on each m² of horizontal
surface, as compared with values of 5-6 GJ m⁻² for parts of the U.S.A.
and Australia. It is worth pointing out that the total national energy
requirement in Great Britain is less than 1.5 per cent of the solar
radiation falling on its surface. About 50 per cent of the insolation is
diffuse rather than direct radiation, and in winter months the greater
part of the solar radiation is in this form, as shown in Figure 3.

It may be also seen from this figure that a south-facing inclined
surface receives more annual solar radiation than does a horizontal
surface, although the horizontal surface receives a higher maximum
figure during the summer months. However, as the heating is
particularly required during the winter months if it is to supplement a domestic hot water system, it is the inclined surface which
is to be preferred to the horizontal surface.

The efficiency with which any solar collector will operate depends very much upon the angle at which it is orientated with
respect to the sun at any particular time, and the ideal situation is
to have the collector at right angles to the sun's rays all the time.
As Figure 4 shows, it is in this position that maximum insolation is
collected whilst in any other position the quantity of solar radiation
striking the collector is reduced. Mathematically this is the original
amount multiplied by cosine Θ, where Θ is the angle between the
horizontal plate and the one perpendicular to the sun's rays.

The need to obtain the maximum radiation possible is of particular importance in winter time when, due to the low elevation of
the sun in the sky, its radiation has to penetrate a greater distance
through the atmosphere.

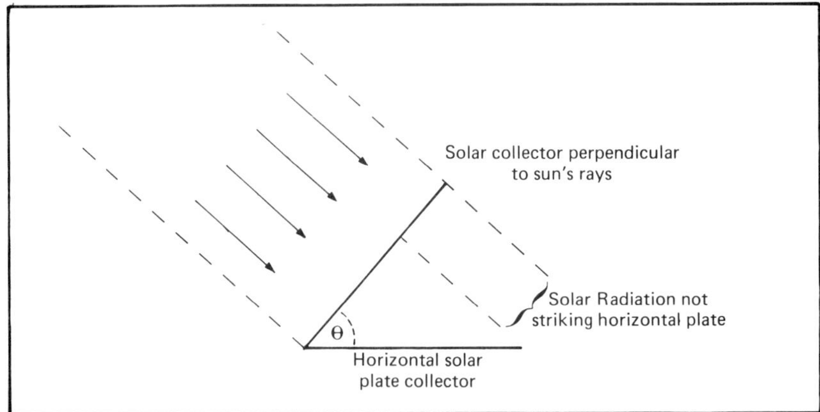

Figure 4.   *Effect of Collector Angle on Amount of Incident Solar Radiation*

36

As the earth is rotating throughout the day and the height of the sun is variable, it would also be desirable for the solar collector to move in phase with the sun in order to keep the collector perpendicular with respect to the rediation. This is actually done with very expensive solar instruments but is prohibitively expensive for anything connected with domestic heating. For these reasons it is usual to have permanently fixed collectors positioned at an angle that will enable the maximum energy to be obtained throughout the year. This optimum angle depends upon the latitude of the building on which the solar unit is to be installed, and the time of year for which maximum insolation is required. Obviously collectors in the northern hemisphere should in any case ideally face due south and those in the southern hemisphere should face due north. The actual optimum angle to the horizontal at which the collectors should be tilted has to be individually calculated and does not necessarily correspond to the maximum elevation of the sun during either a particular day or month. As a rule of thumb, for maximum winter insolation the flat-plate collector panels should have an angle from the horizontal approximately equal to the latitude of the site. However, if allowance is made for the fact that in addition to direct radiation falling on the panel there will also be some diffuse radiation and some reflected radiation from the surroundings, then the true optimum winter angle may actually be 5° less than the latitude angle. In Great Britain a tilt angle between 45° and 50° seems most suitable, and in the U.S.A. between 30° and 40° is applicable.

Where the maximum use of the solar power is during the summer months, for example for air conditioning or swimming pool heating, the angle of tilt can be $10^\circ - 15^\circ$ less than the latitude in order to make use of the sun at higher elevation. In practice, of course, most fixed collectros are attached to established structures where the angle of the collector is predetermined. In these instances it is necessary to use a surface as near the correct angle as possible. Most installers are limited to placing their collectors in pitched roofs where the existing angle is between 30 and 60°. As most people will only have a limited number of locations (or often only one location) for their solar collector, it is the size and efficiency of the collector which is their major concern.

The operation of most conventional solar collectors is dependent upon what is known as the 'greenhouse' effect. This has already been referred to in connection with the heating that takes place inside a car with all its windows closed. It occurs because the solar radiation passing through the glass heats the interior of the vehicle, but the radiation emitted by the heated surfaces is of longer wavelengths which cannot escape back through the glass. As a result there is a build-up of heat within any closed glazed container exposed to the sun. This effect can also be observed in rooms which face the sun if the doors and windows are kept shut. The phenomenon is one that has been used to good purpose by horticulturists ever since the greenhouse was first introduced by Tiberius for growing cucumbers. Loss of heat does occur by other means, which is why there is not a continuous increase in temperature. These heat losses occur due to the three well-known processes of conduction, convection and radiation, and have to be minimised in solar collectors to achieve maximum operating efficiency. The only place at which good conduction of heat is required is between the collector surface itself and the water or air being circulated through the collector. External conduction losses are reduced by constructing solar collector casings from substances that do not conduct heat and by insulating them from the lower external temperatures. Convection losses are minimised by keeping free air spaces as small as possible and radiation losses are reduced by employing special heat-absorbing surfaces and glass or plastic sheets designed to retain the energy wavelengths responsible for the heating effects. It is in this region of improving materials for solar collectors that much research is now being done.

As Figure 1 (page 23) shows, the greater proportion (more than 95 per cent) of solar radiation possesses wavelengths of less than 3 μm. Even ordinary glass is capable of transmitting 80-90 per cent of this radiation and causing the temperature to increase inside any closed container. The energy radiated by the surfaces inside is however greater than 3 μm and this energy is unable to escape through the glass or plastic surface. Heat losses can be further reduced by using two glass or plastic surfaces separated from each other by a small air gap. However, this use of double-glazing also causes the amount of radiation actually entering the collector to go down to about 60-70 per cent of the radiation which strikes it. As a result double-glazed collectors are not normally considered to have any great advantages over single-glazed units.

Another method that may be used to enhance the effect of the solar radiation is the use of flat-plate and curved-surface reflectors. Obviously a series of mirrors all arranged to direct the sun's rays onto a single area will lead to a greater temperature increase than will be obtained by exposing the area by itself to the sun. Even with conventional plate solar collectors it is sometimes possible to increase the amount of energy striking the collector by siting it

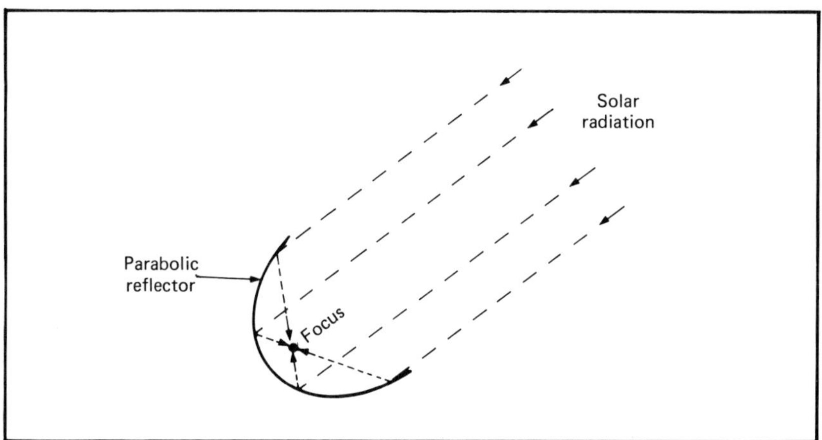

Figure 5.   Focusing Effect of Parabolic Concentrators

near to a white-painted wall or similar surface that will reflect additional energy onto the collector.

The types of concentrators that have been used most extensively have been parabolic mirrors. These have been particularly employed for those purposes in which a concentration of radiation is required at a single point or over a small area. Because of the nature of the parabola, if it is directly aligned with the sun's rays, then reflection of the rays on the parabola surface means that they are brought to a focus at a single point (as illustrated in Figure 5). This is the focal point of the parabola, and is a position that can be calculated mathematically with considerable accuracy. As a result, concentration can be achieved in which the radiation is enhanced by a factor of about 1,000, and even relatively simple designs will give an enhancement of 100 to 200. Such parabolic concentrators do, however, only operate in direct sunlight and not with diffuse radiation. Because of this they are considered unsuitable for wide application in areas in which the number of hours of direct sunlight is less than an average of four hours per day. Another disadvantage is that the reflector rapidly loses its effect if it is not completely facing the sun. This means that for the best results it must be constantly realigned with the sun and either moved constantly in phase with the sun or readjusted at short regular intervals. Another disadvantage of parabolic concentrators is that only one side of an object can be heated, as the other surfaces are turned away from the focused rays.

These disadvantages have not prevented parabolic reflectors being used in solar cookers and for irrigation schemes, although in temperate climates they are more of academic interest than anything else, and the most successful applications in other parts of the world have been large-scale devices involving automatic tracking systems. It is only by the use of direct sunlight on focusing collectors that temperatures exceeding $120°C$ can be attained, and conventional flat solar collectors are unable to attain temperatures even close to this. Because of this, the normal flat-plate collectors are most suitable for providing heating for hot water supplies rather than for central heating systems.

Because of its universal availability, there is no part of the inhabited world that cannot in some way employ solar radiation to advantage. Even areas lacking direct sunlight can use diffuse light, and in many instances the peak quantity of solar radiation is of less importance than constant availability. Applications of solar energy are varied and already widespread throughout the world. They will, without any doubt, increase greatly in the future.

3

USING
SOLAR ENERGY

We all use solar energy already. We use it for drying washing, for drying our cars, for growing our plants, for sunbathing and to a certain extent for keeping our houses warm when it shines through our windows. So the present preoccupation with solar energy is really concerned with making it work more efficiently for us in order to reduce our dependence upon other energy sources. It is worth bearing in mind that when we burn fossil fuels we are really using a stored form of solar energy, as it has been solar energy that has been used in making the plants and animals which have decayed to give the coal and the oil. However, it is stored solar energy which has taken millions of years to accumulate and is being used up in a very small period of time. Moreover, the rate of replenishment is so slow that it can be totally discounted in any energy calculations. Now, as the supplies of fossil fuels are being depleted, we are looking again at the original source of that energy.

Practical applications of solar energy fall into three main groups:

> low temperature application — particularly for heating swimming pools in which large-area unglazed collectors are most commonly employed, and simple heated-air floor systems;

> medium temperature — associated with the use of glazed collector panels for providing home heating, water purification, and for driving solar engines;

> high temperature — those systems using energy-focusing devices for cooking, high-pressure steam generation and solar furnaces.

The simplest and most straightforward use of solar energy is by the gardener with his greenhouse. This is a direct use of the

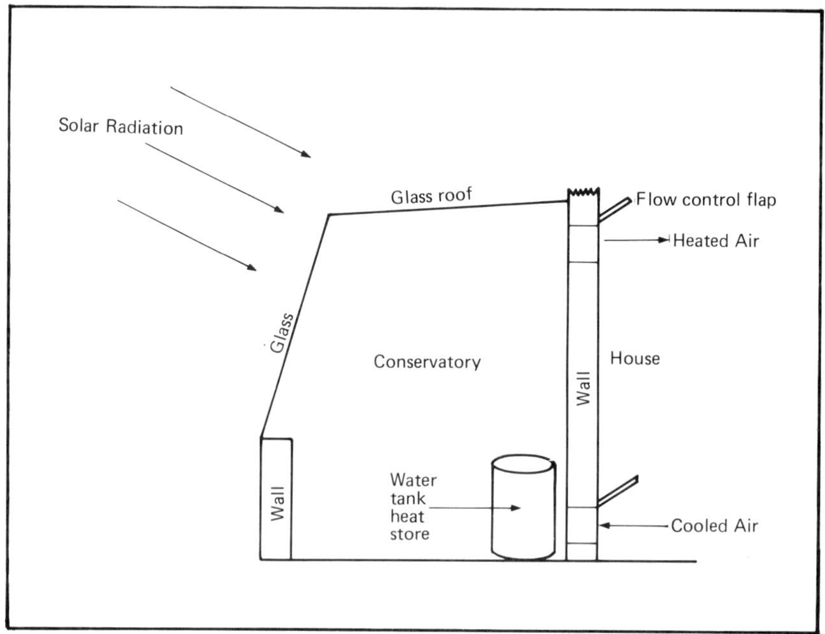

Figure 6.   The 'Greenhouse' effect used to assist house heating

temperature increase inside the glass structure which arises as a result of the inability of the longer wavelengths of radiation to escape through the glass. From a domestic point of view the 'greenhouse' effect still serves as a method of adding to and retaining the heat within a house. Although the old-fashioned conservatory as a lean-to on the back of a house has now given way to the modern sun-lounge extension, it functions in the same way. As the temperature in the sun lounge builds up during the day it can be shut off from the room to which it is joined, but at night time as the external temperature falls, so the stored heat from the sun lounge can be allowed to circulate into the adjoining room. In some experimental cases the solar radiation in the lean-to has been used to heat up containers of water, acting as heat storage systems, so that circulation of cooler air over them later in the day leads to an exchange of heat supplementing that in the rest of the house. These processes are illustrated in Figure 6. Such a system which does not use any form of forced circulation pumps is known as a passive heating system.

One of the most well-known applications of the use of passive solar heating was in the construction of the annexe of the St.

42

Plate II.   St. George's Middle School, Wallasey

George's Middle School, Wallasey, Merseyside (Plate II). The building, which is actually a major part of the school, was designed to provide an environmentally balanced atmosphere without the use of traditional heating systems. To achieve this it is very well insulated, constructed from heavy heat-retaining building materials and with a special double-glazed solar wall made from large glass panels 61 cm (2 ft) apart.

Heating in the building is adequately provided for from a combination of solar radiation, the electric lights (particularly in winter months), and the heat produced by the pupils in the classrooms. During winter months about 25 per cent of the heating comes from solar radiation. The building includes classrooms, offices, laboratory, assembly hall and gymnasium, all of which obtain satisfactory heating at all times.

Although this was a pioneering development, it has not only been highly successful, but has also provided useful data on the type of performance that can be expected from passive heating systems. Although completed as long ago as 1961, the school still receives many visitors wishing to study the design, and its value as a solar building continues to promote much discussion.

Another passive solar heating system has been developed in France by Professor Felix Trombe who has been in the forefront of solar energy developments for many years. The principle of this is similar to that of the adobe houses used by the Pueblo Indians. These had thick walls and small windows to keep interiors cool during the hot days, during which they acquired sufficient heat to act as a radiator and provide internal warmth during the cooler evenings. In the Trombe wall design the essential features are a large double-glazed window facing the sun and a massive internal wall with controllable ventilators built inside the house a short

43

distance behind the window. The successful operation of the system is based upon the principle of hot air rising and cold air sinking, causing a circulation within the house. In winter the wall acts as a heat collector, hot air rising between the wall and the glass panel. During the summer daytime the house is kept cool by allowing air to pass into the house from the cool side of the building and to pass up the front of the internal wall and out through a vent in the window. During the cooler summer evening the house can be kept warm by altering the vents to give the winter circulation system.

One of the first major applications of the Trombe wall principle has been in the construction of nine passive solar-heated homes for elderly people at Acorn Close, Higher Bebington, Merseyside (see Plate III). The scheme, which is the result of collaboration between the British Government, industry and architects, was opened in November 1978 and uses a massive south-facing wall separated by an air cavity from an outer glass wall. The wall acts as a heat store for the solar radiation during the day, and this is released to the air and ducted around the houses by natural convection at other times (Figure 7). It has been estimated that this design will produce a saving of 30-60 per cent in the normal energy requirements for space heating. The whole design is partially experimental and the computer-monitored results from these houses will be compared with those obtained from similar but conventionally heated houses.

Plate III.   Acorn Close, Higher Bebington

44

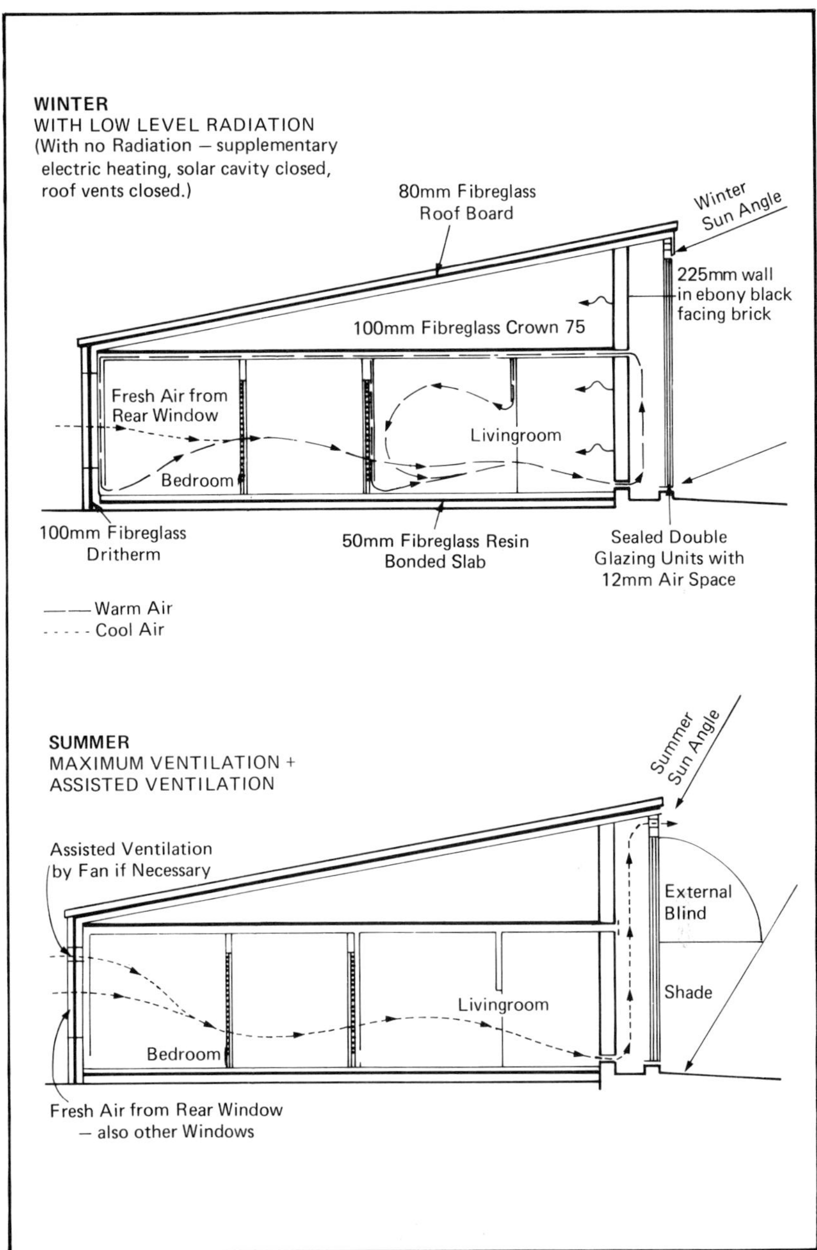

Figure 7.   Air Circulation Diagram for the Passive Solar Heating System at Acorn Close, Higher Bebington.

Reproduced with permission from data provided by Pilkington Brothers Ltd.

Very few people in established houses will be interested in installing Trombe walls, but this does not mean they cannot try out a passive heating system. A very simple approach is to build a glass-fronted box tube at an angle of 45-50° from the ground up to the bottom of a sash window (or a specially constructed window frame). If air can enter the lower end of the tube it will become heated under the glass due to solar radiation and will rise to circulate and heat the room (Figure 8).

In some special solar houses air-ducted systems such as this have been developed on a larger scale in order to take the heat from water, chemical or gravel heat stores to supply the whole house. Most of these are experimental.

For most people the major question is 'How is solar energy to be used to provide hot water?' In most countries it is as a supplement to the hot water supply that solar energy is likely to be most used, as it is unlikely that most systems will provide sufficient heat at the required time to add effectively to central heating without investment in very expensive heat storage systems. This possibility is being investigated, though.

It is generally reckoned that on the average only about one-third of the solar energy falling upon a collector can be converted to heat. In practice the efficiency of the collector varies with the time of year, and to provide sufficient hot water for the average family using about 50 gallons every day it is necessary to have a collector surface covering about 6 m² (65 ft²). This heat collection is carried out using a flat-plate collector which has the function of raising the temperature of a heat transfer liquid (usually water) passing through. Most collectors are fitted into indirect heating

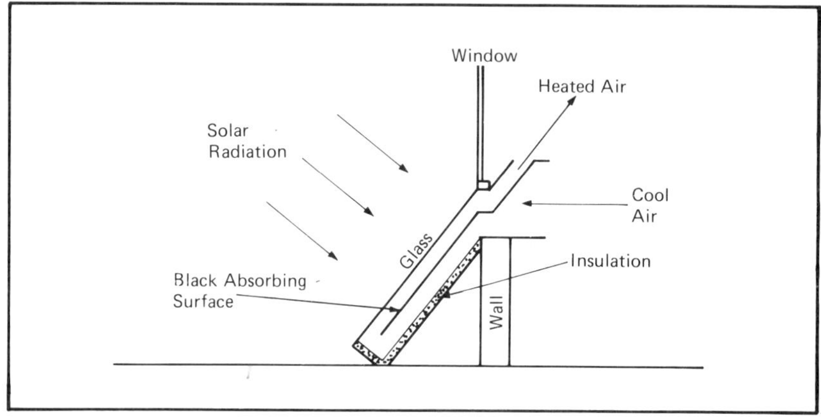

Figure 8.   Passive heating from a window frame unit

46

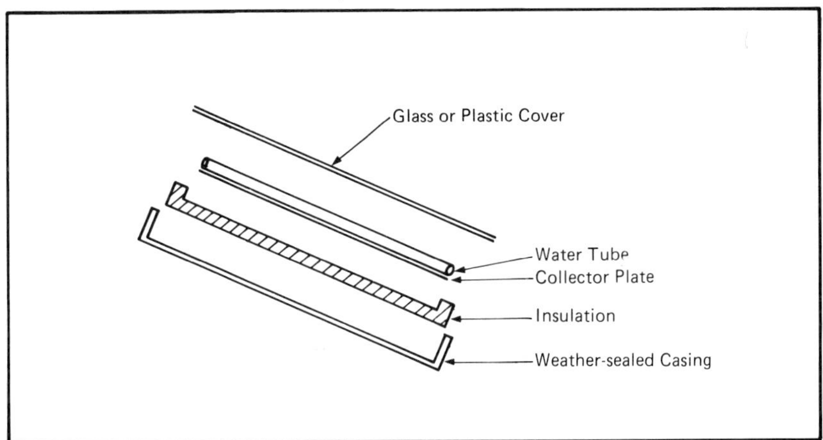

Figure 9.   Main design features of a flat-plate collector

systems so that the heat from the solar panel is transferred to water in the main system hot tank.

The first important point is the design of the flat-plate collector: it consists of a transparent cover, an absorption plate, a liquid flow system, insulation and casing. The essential construction of a flat-plate collector is shown in Figure 9.

The cover is most commonly glass, although various plastics and plastic-reinforced glass have also been developed for this purpose. Glass is not transparent to ultra-violet radiation and about 10-20 per cent of the arriving solar radiation is lost due to reflections at the air/glass interfaces and absorption within the glass itself. In terms of the overall transmission of solar energy there is little difference between glass and the better plastics such as polyvinyl chloride. Glass is, of course, more liable to breakage, whilst plastics are cheaper but tend to become discoloured over a period of time, leading to a reduction in transmission. The glazed cover should be capable of transmitting 80-90 per cent of the radiation which reaches it, and periodic cleaning to remove accumulated dirt and leaves may be necessary to prevent a fall-off in performance. It is usual to use single sheets of glazing although some panels are produced double-glazed, in which case only 60-70 per cent transmission can be expected. As the use of double glazing leads to this reduction in the amount of radiation penetrating to the absorption plate, it is doubtful if the additional cost is justified despite the improvement in preventing heat losses due to re-radiated energy. The radiation absorption surface and liquid flow system are usually integrated into a single unit in which the purpose is to absorb the

47

heat and transfer it to the water as efficiently as possible. Maximum absorption is achieved by black surfaces, which only re-radiate energy of longer wavelengths that cannot pass out through the glazed surface. The absorption surface may be a black plastic sheet in close contact with the tubes containing the heat transfer liquid. But the best absorption is obtained with specially prepared surfaces which can produce an absorption efficiency of 80-100 per cent so long as there is good thermal contact between the absorbing surface and the flowing liquid.

In its simplest form the surface can be a coat of matt black paint, but the more highly selective absorbant surfaces consist of electro-deposited films such as black nickel (a nickel sulphide/zinc sulphide complex on a nickel base) or black chrome (a chromium/chromium oxide complex on a nickel base). Black copper oxide on copper and aluminium surfaces has also been used, and selective coloured coatings on stainless steel have been produced by immersing the metal in hot aqueous acidic solutions. As a result blue and purple flat-plate collectors are now produced in Japan and Australia in addition to the conventional black ones.

It is, of course, essential that such surfaces are cheap to produce in order to keep the overall cost of production of the solar collector as low as possible. At the same time the surface must have a high durability, able to withstand substantial variations in temperature in addition to the effects of atmospheric pollutants and corrosion.

Although the most common domestic solar collector systems use water as the heat transfer fluid, it is possible to employ flat-plate collectors using air. These have been tried in Florida and in parts of Australia to supplement the space heating in the rooms of houses. However, this is normally operated in conjunction with a heat-storage system using a large volume of rock, gravel or water and requiring an air-ducting system down to the storage space and back around the house. The obvious advantages with the use of air is that it cannot freeze and if leaks arise they do not cause any serious problems. At the same time it is not as good a medium for heat transfer, as it lacks the high density and specific heat of the liquid systems.

Water has many virtues as a solar collector fluid, not least of which is its cheapness. However, as there is always the possibility of freezing in the winter months and of corrosion setting up within the metal tubing, it is frequently necessary to add anti-freeze based on ethylene glycol and anti-corrosion agents to the water which then has to circulate in a closed system, the heat from the solar collector being passed through a heat exchange coil in the hot water storage tank. Adequate protection against freezing can be achieved with a concentration of 15-25 per cent of anti-freeze in the circulating water. Where such additives are not used, the water flow can be used as part of a direct hot water supply system in which the water actually heated in the solar collector is used for domestic purposes, rather than in the indirect systems in which its heat is exchanged through a heat exchanger coil to the domestic hot water in a water tank. These are discussed in more detail later in this chapter.

The design of the water flow pipes is of considerable importance, and normally only three metals are employed for this purpose: copper, stainless steel and aluminium. It is desirable to obtain the maximum possible heat transfer to the tubing and hence to the water in the tubing. The efficiency with which different materials achieve this depends upon their thermal conductivity — this is given the symbol 'k' and is a measure of the rate at which heat is transferred across a material. The 'k' values for a number of common metals are given in the following table:

silver	4.17 watts $\mathrm{cm}^{-1}\,\mathrm{K}^{-1}$
copper	3.82
gold	3.10
aluminium	2.30
nickel	0.83
chromium	0.85
stainless steel	0.25

This shows quite clearly that of the non-precious metals, copper is by far the most desirable for this purpose. The heat transfer is also assisted by using tubes with as thin walls as possible. Copper tubing of outside diameter 15-28 mm (0.6-1.1 inch) and with a wall thickness of between 0.6-0.9 mm (0.024-0.035 inch) is commonly employed. Under normal conditions this has a working life expectancy in excess of twenty-five years. Where mixed metal systems are used, as might be the case for an aluminium collector panel with stainless steel tubing and copper hot water tanks, precautions against electrolytic corrosion are essential. These can take the form of corrosion inhibitors in the circulating water and a sacrificial aluminium anode which will corrode in preference to the collector panels.

Various designs and arrangements of tubes have been employed in solar panels. One of the cheapest systems has been a do-it-yourself solar water heater based upon the use of second-hand panel radiators, devised by the National Centre for Alternative Technology,

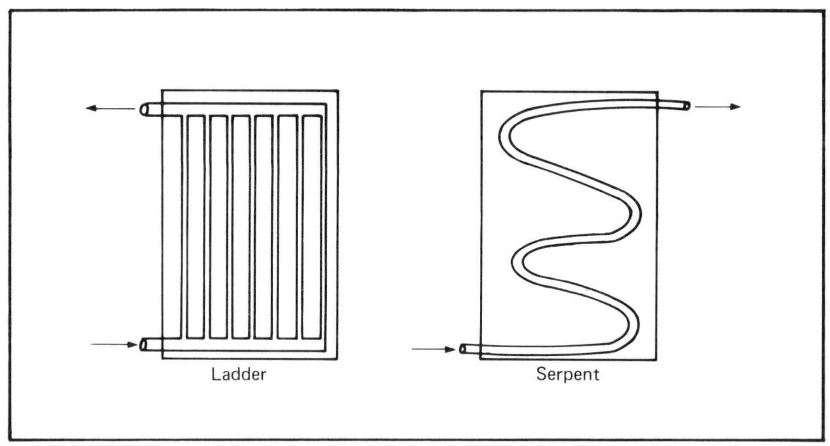

Figure 10.   Tube layouts for flat-plate collectors

Machynlleth, Powys, Wales. But the most common commercial methods are to use either a ladder structure or a S-shape of copper tubing (Figure 10) for each panel, with the water flow being from the bottom to the top.

Where several panels are used together on a roof flowing from one to another, careful arrangement of the panels relative to each other is important to avoid any dead spaces. In addition to this the tube arrangement must be such that it is possible to drain the system down without leaving any undrained sections as would occur, for instance, if the zig-zag design was placed sideways. The desirable arrangements for joining several panels are shown in Figure 11.

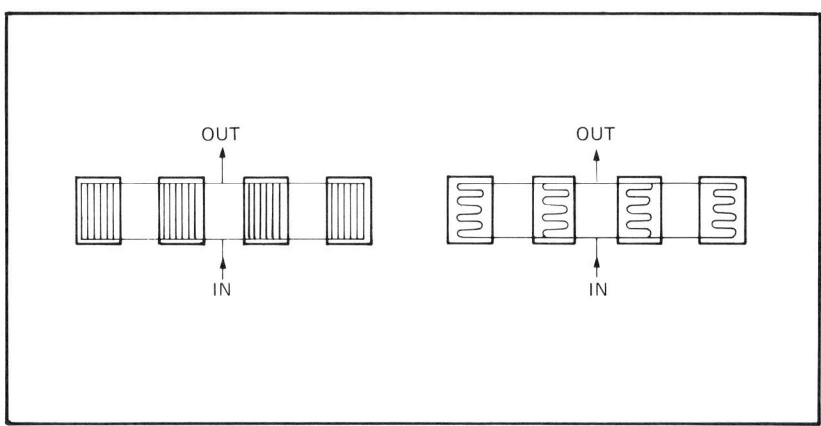

Figure 11.   Roof arrangements for flat-plate collectors

50

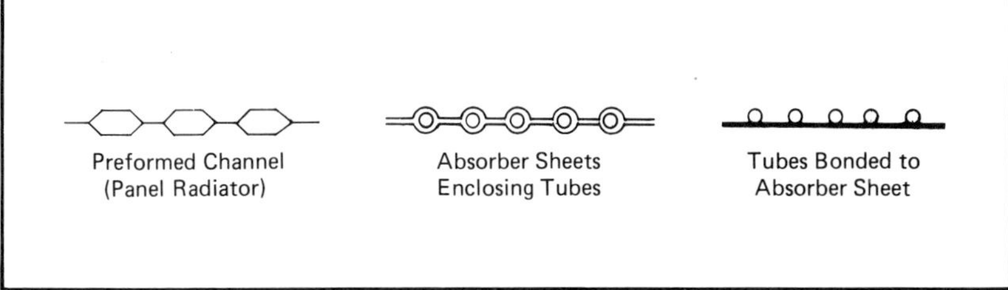

Figure 12.   Cross-sections of tube and absorber designs

Where a separate tubing system is used it is necessary to bond this by soldering, brazing or welding to the absorption surface in order to maximise the heat transfer. This means that a variety of combinations of tubes and absorption sheets are in use. The cross-sections of some of these are illustrated in Figure 12.

Having achieved a good heat transfer system it is important that any heat collected within the collector should not be allowed to escape, so insulation between the collector elements and the outer casing is of paramount importance. For this purpose a minimum of three inches of fibreglass wool or polyurethane foam is required, and ideally there should be a layer of a highly reflective material such as aluminium foil between the fibreglass and the bottom of the absorber plate. This serves to reflect back any radiation emitted from that surface of the absorber. Insulation at the sides of the panel is also important.

A typical size for a flat-plate collector panel is between 1.5 and 2.0 m^2 (16-21 ft^2) with an overall thickness of about 0.1 m (4 inches), and weighing about 25-35 kg (55-77 lb) for every square metre of collector. Such collectors can be readily fitted onto an appropriate inclined roof, or even set into the roof like a fanlight. Alternative locations that have been successfully used include the roofs of garages, single storey lean-tos and even the pitched roof over porches. Where there is no alternative method available, as with flat-roofed houses, the panels have been mounted on specially constructed frames. In these cases the frames can be designed to provide the optimum inclination and positioned to obtain the maximum exposure to the radiation. Typical arrangements of solar panels installed on established roofs are shown in Plate IV.

Another important feature about the installation of the solar panels is that all connecting pipes flowing to and from the panel must be very well insulated. Without adequate insulation it is possible to lose a very large proportion of the solar heat that has been transferred to the water.

51

Fairseat, Kent

Basing, Hampshire

Sevenoaks, Kent

Plate IV.   Solar roof panels   *Southwark, London*

52

A number of different flow designs have been drawn up enabling solar panels to be linked in with established domestic water systems. In the simplest of these the circulation of the solar-heated water is dependent upon the basic principle that hot water rises and cold water sinks because of the difference in the specific gravity of the water at different temperatures. This is known as a thermosyphoning system and necessitates the solar panel being situated at a level at least 30 cm (1 ft) below that of the water storage tanks. The essential layout required for this approach is shown in Figure 13. In this thermosyphoning system the hot water from the panel(s) may be fed direct to the tank from which the domestic water is drawn off. This prevents any anti-freeze or anti-corrosion agents being added to the water.

However, there is a great advantage in fitting a small pump into the system to circulate the solar-heated water, as it enables smaller bore tubing to be employed and the panels to be mounted at a higher elevation relative to the water tanks. By using a pumped system, too, it is possible to integrate the control of the solar heating into the overall water demand, so that by the use of thermostatic controls the water will only be circulated through the panels when the panel temperature is above that of the water

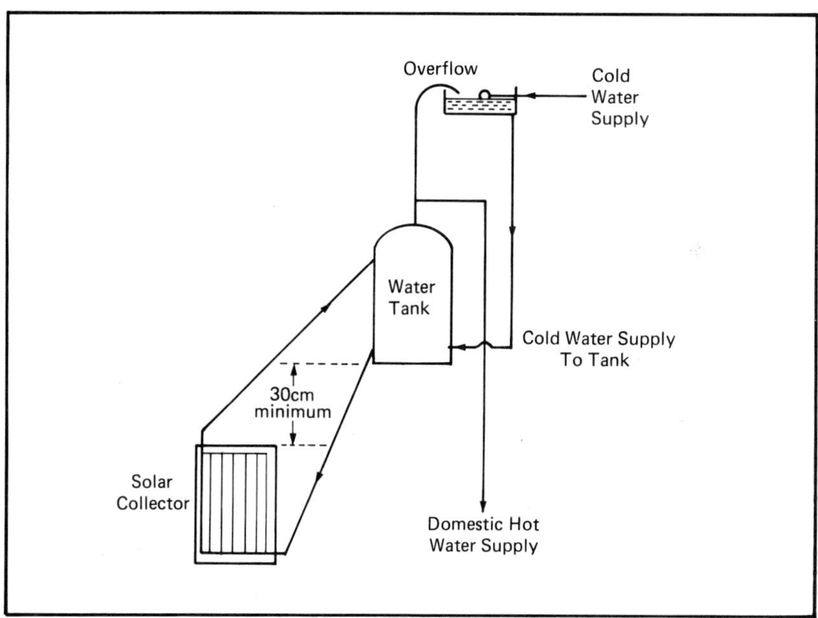

Figure 13.   Circulation for Thermosyphoning Domestic Solar Water Heating System

Plate V.   Solar panels on staff cottages at the Centre for Alternative Technology, Machynlleth, Wales

tank and a net gain in heat can be anticipated. Even the pump can be solar-powered if required, by using a solar cell panel (see Chapter 5) to charge a storage battery. A total solar system like this has been constructed at the Centre for Alternative Technology, Machynlleth, Powys, and employs a 30W circulating pump which takes its power from a solar-charged 6 kWh battery. The combination of roof and ground level panels used on staff cottages at Machynlleth is shown in Plate V.

Pump systems are usually designed to switch in the solar panel flow when the panel temperature is 4-5°C above that of the water tank. This ensures that heat is added to the water tank, not taken from it. The pump and thermostatic control also make it easier to safeguard the whole water system against freezing. The type of arrangement employed in a pumped system is illustrated in Figure 14.

The number of solar panels that are needed will, of course, depend upon the size of the property, the number of people for whom the water is required and the amount of the heating bill which it is intended to eliminate. The efficiency with which the panels will operate will also depend upon them being carefully

54

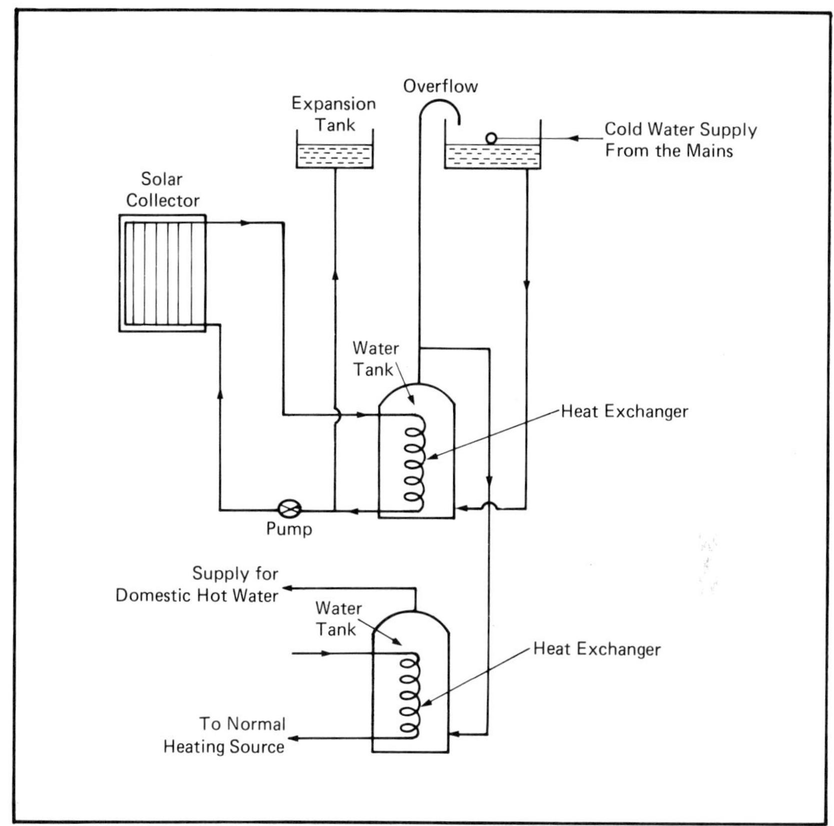

Figure 14.   Circulation for pumped solar domestic water heating system

maintained and the transparent cover being kept clean. A rapid fall-off in the amount of energy collected occurs once dirt starts to accumulate on the surface. In Great Britain it is considered that solar heating can provide up to 40 per cent of the annual hot water requirements for the average home, with a larger proportion (90 per cent) being provided in the summer months and a lesser proportion (about 20-25 per cent) in the winter. An average household requires heat for about 230 litres (50 gallons) of water with the temperature to be raised from about 10°C (50°F) to 60°C (140°F). In most instances a collector area of 6 m² will be capable of providing almost all the requirements on a hot summer's day, but will require supplementary heating on cooler days and particularly during the winter. In many instances the installation of a larger collector area, in order to provide a larger proportion of the winter heating, results in over-provision for the summer months and is

Plate VI.   Wates houses at Crofters Mead, Forestdale, Croydon

Plate VII.   Students' flats at Llys Tal-y-Bont, Cardiff, Wales

56

considered to be less cost-effective. As domestic water heating absorbs just over twenty per cent of energy costs in the home, it is apparent that solar water heating can lead to a reduction of 5 per cent in the annual energy bill.

Solar collector systems of this type have aroused a great deal of interest and many private installations have been made. Major building contractors have considered incorporating them in new houses, and one of the best designs has been a group of three houses (Plate VI) with integrated solar collectors built by Wates as part of a larger estate at Crofters Mead, Forestdale, Croydon. The actual incorporation of the collectors as part of the roof meant that their cost was partially offset by savings on other roofing materials. The long-term results from these houses may serve as a justification for building more homes of this type.

A larger application which is being studied in depth is that for the provision of hot water to student flats at Llys Tal-y-Bont, Cardiff (Plate VII). These have been in operation since mid-1977 and use a direct water circulation system without an intermediate heat exchanger, so that the actual solar-heated water is used for the hot water supply (Figure 15). Supplementary heating, for when the solar energy is inadequate, is provided by an electric immersion

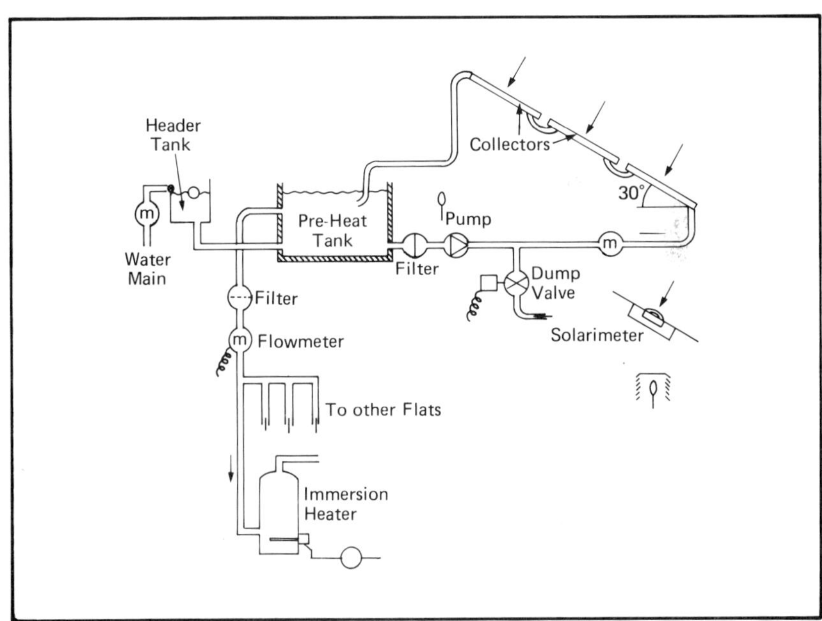

Figure 15.   Circulation system for solar heating system at the Students' Flats, Llys Tal-y-Bont, Cardiff, Wales.

heater. The solar collectors, angled at 30° to the horizontal, have a total area of 25 m² (270 ft²) and provide heat to an 800 litre (176 gallon) preheat water tank supplying a total of twenty students in four flats below. As this is a direct system to which anti-freeze agents cannot be added, it has been constructed with an automatic control for draining the water from the panels when the panel temperature approaches freezing. Data obtained from this installation is automatically recorded to provide an accurate assessment of the energy gained by the system, and is compared with that used in the other flats with conventional hot water supplies. The installation has been funded from University College, Cardiff, and was designed by the Solar Energy Unit under the direction of Professor Brinkworth.

Not all installations are experimental, however, and many people who have made a private investment in solar collectors speak very highly of the results. One of the largest installations in Great Britain must be that at the Mountain Hotel a few miles south of Brecon, Powys. There three rows of solar panels (Plate VIII) covering an area of 52 m² (560 ft²) are used to provide hot water, mainly for the bedrooms in the hotel. The average daily hours of sunshine vary from 0.92 in December to 6.14 in April. With the additional contribution due to diffuse radiation, more hot water is obtained than is required for much of the year. Although the system has not been studied from a scientific or academic point of view, the

Plate VIII.   Solar panels on the roof of the Mountain Hotel, Brecon, Wales

58

hotel manager, Mr. Vernon Phillips, is satisfied that his solar collectors make a major contribution to keeping his energy bills down and is very happy with the results from the installation, which has been in operation since early 1977 and has received much interest from the hotel and catering trade.

There is no such thing as a typical set of results from solar collectors. Not only do panel designs differ, but house positions vary and the water usage of families is probably the greatest variable of the lot. However, the following figures were reported from 5 m^2 of south-facing collectors, at 45° to the horizontal using a pumped circulating mixture of water and ethylene glycol (anti-freeze). Monitored over a two-year period, useful performance was obtained in every month of the year, with the panels providing heat for an average of 1.7 hours per day in January and up to 7-9 hours per day in July. The average over a whole year was 4.4-4.5 hours per day of useful heating. After allowing for depreciation of the system over 20 years, it was calculated that the cost of energy saved corresponded to an investment return of 16-18 per cent on the capital employed.

Another, smaller system, using 4.3 m^2 of collectors facing 15°W of South at 50°30' to the horizontal was similarly calculated to give an 11 per cent return on investment after allowing for depreciation over twenty years.

In all of these systems using solar panels with circulating water, the intention has only been to provide domestic hot water. This is mainly because to use solar radiation for space heating in Great Britain requires a large collector area, which is not necessary in summer, and some large-capacity heat-storage system. Most of the studies on heat storage have been made in the U.S.A., and have used special cellars below the solar houses which have been filled with stones, water or special reversible thermal chemical processes. These operate by having any surplus heat from the solar collectors passed into them either by heat exchangers or by direct water circulation, and then the heat is extracted and employed for space heating the rooms in the house as required. The success of such methods depends upon having a well-insulated heat store and a large enough volume and sufficient heat capacity to provide space heating even if there is no available solar radiation for several days.

One major study involving heat storage now under way is the first fully monitored solar house to be built in Great Britain which uses solar radiation for space heating. This was designed by a research group at the Polytechnic of Central London under the supervision of S.V. Szokolay, and the system was built into a

Plate IX.   Milton Keynes Solar House

timber-frame house constructed in 1975 by the Development Corporation at Milton Keynes. In this solar house (Plate IX) the heat storage uses two large water tanks with a total capacity of 4.5 m^3 (approx. 1000 gallons), in preference to pebbles, which were considered to be too bulky, and chemical storage, felt to be too complex.

The Milton Keynes house is rented by the Development Corporation to ordinary families who occupy the premises while the energy balance is monitored continuously around them.

Although the occupants are very much technological guinea pigs who tolerantly put up with frequent visiting observers, their only inducement is the possibility of lower energy bills.

The collector area for the house is 37 m^2 (approx. 400 ft^2) contained in the 30° slope of the roof and is intended to provide heat to maintain the heat store at least at 40°C (104°F). Additional heating can be supplied, when required, from a gas boiler and the heat from the store is distributed by a fan convector. Hot water for the washing and for baths is obtained by heat-exchange coils fixed inside the heat store. The basic layout of the solar heating system is shown in Figure 16.

It had been hoped that this would provide up to half of the total heating load, but this has not been realised in practice. However, despite this the whole project is considered to be a great success. It has proved the feasibility of such a project, has shown the limitations of certain items of equipment, and has indicated improvements that can be made in future installations, particularly

60

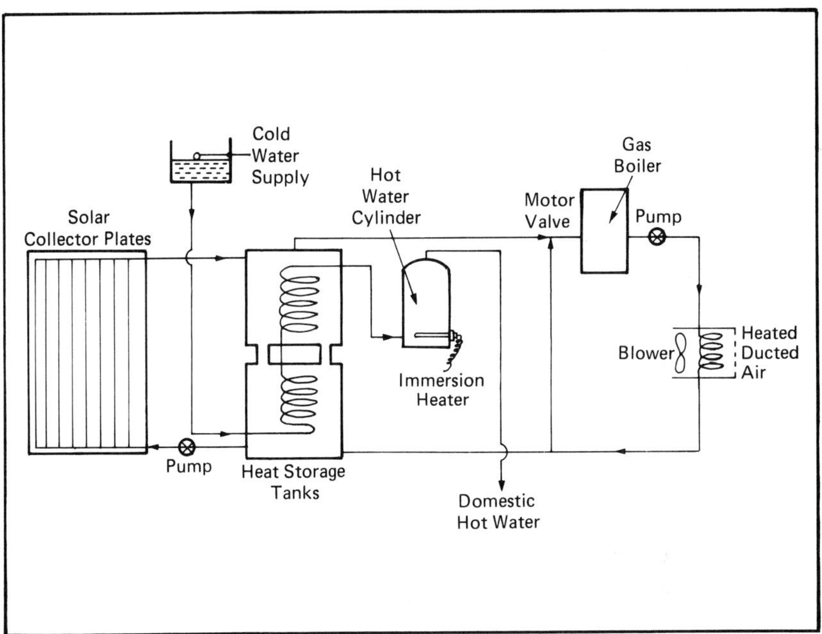

Figure 16. *Water circulation and heat storage system in the Milton Keynes Solar House*

in terms of house insulation. Studies on this house will continue for several years whilst other solar energy schemes are developed in different parts of Milton Keynes.

In hotter climates such as those in Australia, parts of the U.S.A. and Israel it can be shown that there is a clear economic advantage in employing solar collectors for domestic water heating. As things stand at present in Great Britain, there is much conflict in the figures being produced. In 1976 the Building Research Establishment at Garston, Watford produced figures showing that after allowing for depreciation and interest on capital, solar heating was not at that time cost-effective. Another study was made by the Consumers' Association and published in *Handyman Which* in August 1977. This showed that a professionally installed flat-plate solar heating system would cost £650 for the average house and was less cost-effective than other forms of energy conservation — although a successful do-it-yourself system was likely to be financially worthwhile. However, since that date substantial increases in the cost of oil, electricity and gas have exceeded those required to make solar heaters cost-effective, and it is reasonable to say that a system with a lifetime of twenty years will justify

61

its cost if installed by an established company. For a properly installed do-it-yourself solar system, assuming that labour costs are free, the economics of the whole thing become much more on the positive side and less marginal. As energy costs continue to rise the justification for more solar collectors will progressively increase so long as the cost of materials does not rise in line with those for the energy sources the collectors are required to replace.

The interest in solar energy is increasing throughout the world, and leading manufacturers are now looking to markets as far afield as Chile and China (Plate X). But the applications are not limited solely to domestic hot water supplies. Solar power is of particular value in providing hot water for remote areas which are only used on a casual or seasonal basis. In many instances the cost of installing cables or pipes has deterred owners from providing electricity or gas supplies to caravan and camp sites, but the installation of solar panels avoids the expense and regular bills associated with normal energy supplies.

An alternative procedure to using solar panels for providing hot water on a casual basis is what has become known as the 'bread box' type collector. This consists of a 50-100 gallon tank painted

Plate X.   The Duke of Kent visiting the Solarsense Stand at the British Energy Exhibition in Peking in June 1979

62

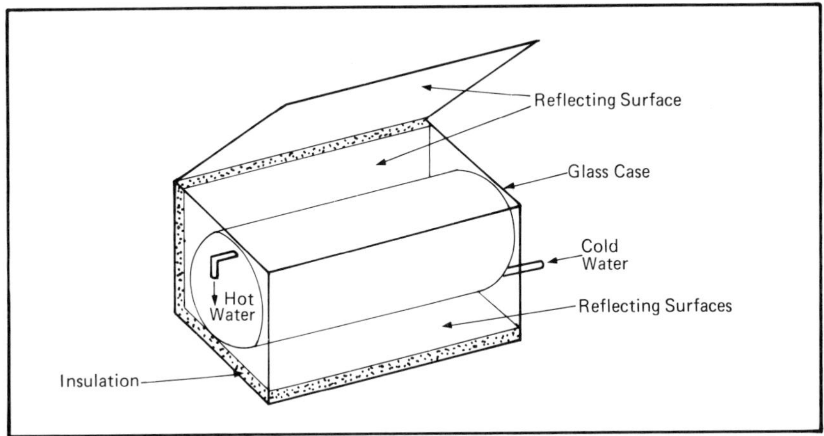

Figure 17.   *'Bread Box' type of solar water heater*

black on the outside and placed inside a specially constructed box with glass panels on two faces and two sides, which are orientated towards the sun. The remaining two sides, by which the tank is supported, are both well insulated and coated with a reflecting surface on the inside. Further solar radiation is passed into the box by using a hinged reflector lid and front, and the completed arrangement is placed in an unshaded area and orientated towards the sun. (Figure 17). Such an arrangement is capable of providing hot water at about 50°C (122°F). Although it has the disadvantage of taking a long time to heat such a large bulk from cold, it has the advantage that it will hold this heat for some hours after the sun has set and still provide warm water into the evening of a hot day.

The average home does not possess a swimming pool in the back garden and many people will dismiss the idea of using solar radiation for heating swimming pools as being totally irrelevant. However, there is an increasing application for this purpose not only for small private open-air pools but also for larger pools owned by schools and local authorities (Plate XI).

Solar collectors for swimming pools are also available commercially and are of a more simple design than those for domestic hot water, as they are only needed to raise the water temperature a few degrees above that of the surroundings, rather than to the higher temperature required for hot baths and washing. For these low-temperature purposes it is possible to use unglazed collectors with large surface areas and a large volume of water flow. By this means even in the temperate climate of Great Britain it is possible to raise the swimming-pool water temperature by 5-10°C

Plate XI.   Solar heating unit for the swimming pool at Ye Olde Felbridge Hotel, East Grinstead, Sussex

(9-18°F) above that of the unheated pool and to extent the use of the pool over a total of six months of the year rather than the more common period of late May to early September, without employing other forms of heating.

Adequate solar heating can be obtained with unglazed heating panels equivalent to about 50-60 per cent of the surface area of the pool. These are usually free-standing on a frame that can be moved around and inclined at 30°-50° from the horizontal. They are sometimes known as open-surface trickle type solar collectors. The water from the swimming pool is pumped to the header tube at the top of the panel and then runs out through small holes down channels on the blackened surface of the absorption plate. Here it picks up the solar radiation and then runs off into a gutter or collector pipe, from where it is fed back to the swimming pool. A typical design is shown in Figure 18.

For this purpose the absorption panel may be made from corrugated metal sheets with black matt surfaces, or ridged-surface black plastic sheeting. All of these have been found to be perfectly effective for this purpose, but costs for the commercial systems vary enormously and simple do-it-yourself systems are not difficult to design. All metals proposed for this purpose suffer from the potential disadvantage of corrosion, as they are exposed to water which is chemically treated and contains chlorinating agents. Direct exposure

64

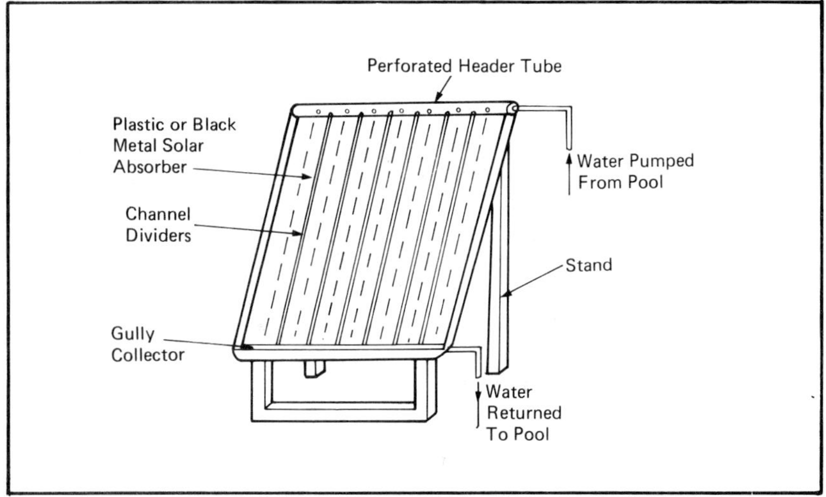

Figure 18.   Solar collector design for swimming pools

to the ultra-violet radiation of the sun through being unglazed also leads to some unpredictable effects. As a result of these problems with metals, there is an increasing use of plastic materials, such as polypropylene, for this purpose.

Although glazed panels are not necessary in order to obtain adequate heating for a pool, a number of commercial glazed systems are available to the public. Some of these employ black matt painted aluminium absorption panels whilst others use black plastic. The main disadvantage with glazed panels are that the angle of orientation is more critical than it is with unglazed panels, and as they attain higher temperatures the plastic ones are more prone to deterioration due to the sun's action if they are left exposed with no water circulating through them. The main advantage of using glazed panels is that a smaller panel area of only 25-30 per cent of the surface area of the pool is required to heat the same volume of water to the same temperature.

Because of the large surface area that is usually employed for the solar heating of swimming pools it is sometimes difficult to find the space for mounting the frames with a good exposure that is not overshadowed by buildings or trees. In these situations alternative methods of capturing the solar radiation are used. One approach has been to use the paved area around the swimming pool as a solar collector with the water circulated through a long black plastic pipe below the paving slabs. This approach is obviously open to a wide range of variations using properly insulated pipes under plastic or glass paving slabs, similar to an ordinary flat-plate roof collector. Solar collector paving slabs covering an area about half that of the pool are normally needed to provide adequate heating. Another alternative method of heating a swimming pool has been described by Dr. J.C. McVeigh, Head of the Department of Mechanical Engineering at Brighton Polytechnic. The system, which has been installed in a swimming pool in Sussex, involves circulating the pool water so that it forms a series of small waterfalls over the steps leading into the pool. By coating the steps with a black surface, and with a sufficiently large area exposed to the sun, sufficient absorption of solar energy is obtained to raise the temperature of the water.

It must also be borne in mind that the water in the swimming pool also serves as a solar collector, and is capable by itself of absorbing 70-80 per cent of the solar radiation that strikes it. It will, however, take several days for the unaided pool to reach a suitable temperature even when exposed to fairly high temperatures. Once the pool has become hot enough for swimming, the solar heating system is only required to compensate for heat losses occurring overnight and during overcast days. At the same time, in order to retain as much heat in the pool as possible, it is desirable to use a cover for when the pool is not in use. Some modern covers allow solar radiation to penetrate and act as partial solar collectors themselves. By means of thermostatted controls it is possible to arrange for the pool water only to be circulated through the solar collector during daylight hours at times when the difference between the water temperature and the surrounding temperature are sufficient to give a temperature gain to the circulating water.

The economics of using solar heating for swimming pools is rather confusing, as it all depends upon the length of time throughout the year for which the heating is required and the area of collector installed. However, the United Kingdom section of the International Solar Energy Society has assessed the costs involved and considers that installation of solar heating for swimming pools in the United Kingdom is worth carrying out as it will pay for itself within a period of less than ten years.

There is, however, more to it than this, as a great deal depends upon the extent of use of the swimming pool. If it is intended that the pool should be used for more than six months in the year then in the British climate it is impossible for adequate heating to be obtained from solar radiation alone and some form of back-up heating is required, particularly for the winter months. However, it has been shown that the introduction of solar heating for swimming pools is justified solely on the basis of the quantity of fuel oil saved

66

over the six months from April to September. Where an oil heating system already exists, therefore, it is still possible to gain financially by using solar energy when possible.

It should be emphasised that this is again an area in which there is little point in installing excessive solar panels in the hope of obtaining adequate heating for the whole year. This means that there is surplus capacity during hot weather and the larger area leads to greater losses from the panels themselves with a corresponding drop in efficiency. As a result the whole exercise becomes more expensive and the economics make it less justified.

There is one main criticism about using solar radiation for heating in the absence of any other form of heating. This is that there is no rapid method for heating the swimming-pool water after it has dropped below temperatures suitable for swimming, and there is no way that solar radiation can cheaply overcome this problem. Even so, a solar-heated pool which is effectively covered when not in use should retain its heat for 2-3 days in the absence of further solar radiation. It should also be pointed out that swimming-pool solar heaters can be used to heat any large exposed volume of water. One increasing application is in fish-farming, where a faster growing rate of the fish can be achieved if the water temperature is kept above normal. Solar panels enable this to be done at very little cost for both young and mature fish as they effectively extend the normal growing period (Plate XII).

Plate XII. Solar panels used for fish-farming at the Centre for Alternative Technology, Machynlleth, Wales

Another application of solar energy which is attracting increasing interest internationally is in helping to produce usable water from saline or brackish water. Within Great Britain there is, fortunately, very rarely any great shortage of drinking water, as this is normally a problem associated with hotter countries. And it is here that solar radiation can provide one of its greatest services. As has been previously mentioned, solar distillation for the large-scale conversion of salt water to drinking water was carried out in Chile during the last century. At present it is common to produce large quantities of distilled water by other methods of heating at a relatively low cost, although as fuel prices rise the viability of solar radiation becomes more established. It is, in any case, already the most economical method for producing small quantities of distilled water, probably up to 100,000 gallons per day. As a result solar stills are much more common than many people imagine. Several stills are in use in Greece, one of the largest at Patmos has a 93,000 ft^2 surface and produces 7,000 gallons of water per day. Solar stills are also working in Australia, Spain, Tunisia and the West Indies, and a wide variety of structures and designs have been tried. Glass-covered stills are generally considered to be superior to plastic-covered ones, and large durable glass stills can produce purified water at a cost of about \$3-\$4 (£1.50-£2.00) per 1,000 gallons.

Allowing for an efficiency of 35 per cent it is possible to distil 0.33 cm^3 of water for every cm^2 of solar-still surface on a day in which 120 J cm^{-2} per day are received. This corresponds to about 1 gallon of water for every 15 ft^2 of surface, although higher levels of production are possible on really hot days. Although this might appear to be low, it should be emphasised that the basic equipment required for solar distillation is cheap to make and the running costs in terms of labour and maintenance are small.

The basic requirements for a solar still are a large, fairly shallow tray or pan with a black surface capable of holding a 2-3 cm depth of water, plastic or glass sheets covering a pitched roof-shape framework covering the tray, transparent ends to the tunnel and two gutters to collect the purified water. A typical design is shown in Figure 19. The whole system is based upon the fact that water does not have to boil to evaporate, and that as its temperature rises so does its vapour pressure: it is this vapour that can be condensed out. The sun's rays pass through the transparent cover and are absorbed by the black surface of the pan. The vapour pressure of the water is increased because the heat is trapped inside the cover as the longer wave infra-red radiation is unable to escape; the water vapour condenses on the inside surface of the cover, and as the

68

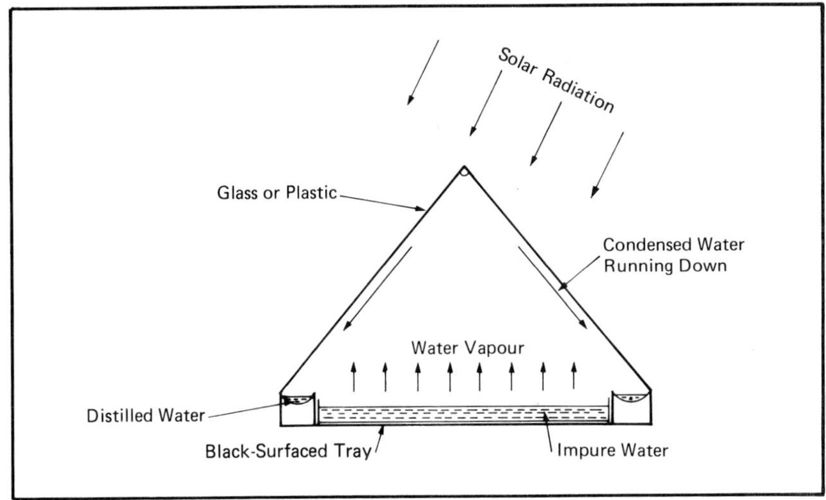

Figure 19.   Principle of solar distillation

droplets build up it runs down the inside of the slope to be collected in the gutters. From there it is conducted to a storage tank.

Another type of solar still can be made from a form of cold frame which has been well insulated all round and fitted with a black-coated tray to contain the brine. It is important in all of these to ensure that the wind cannot get inside the still, as that will cool it down and drive out the water vapour. Even in the absence of a brine tray a cold-frame type of solar still is capable of obtaining water from the earth floor where it is set down. See Figure 20.

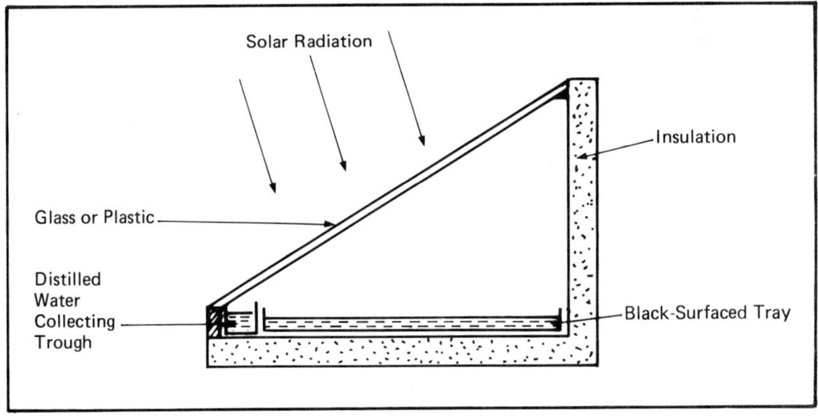

Figure 20.   Cold-frame type solar still

69

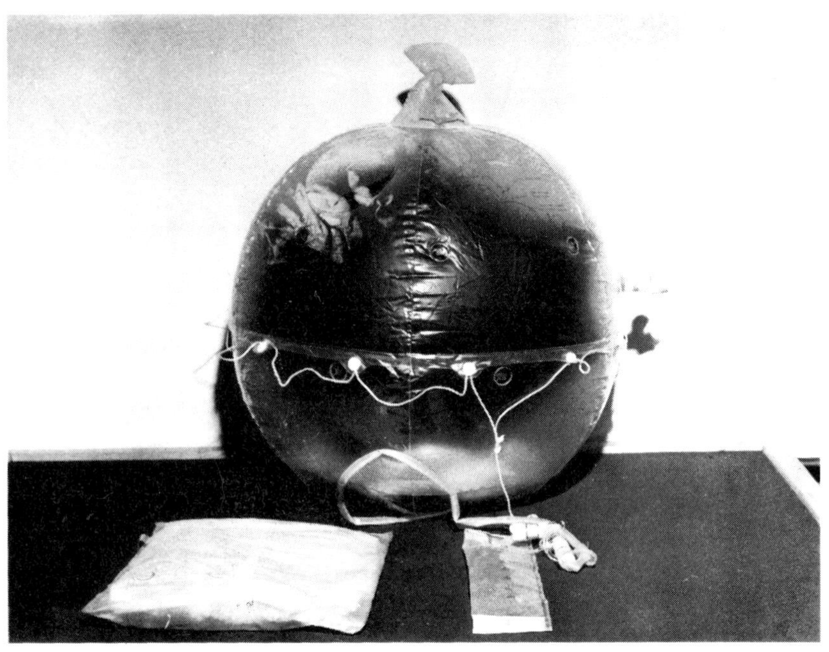

Plate XIII.   *Plastic still used by the armed forces for emergency purposes*

Various forms of inflatable plastic stills have been invented: one is used particularly by the armed forces for people wrecked at sea so that they can purify enough water from the sea to stay alive (Plate XIII). It is also possible to make an emergency still for use in the desert from a sheet of clear plastic, the sand and a cup. For this operation a hole is excavated in the sand and the cup placed in the centre at the bottom. The sheet of plastic is then spread loosely over the hole so that it will hang down in the middle. Sand is placed around the circumference of the plastic to hold it in place and to make it airtight, and a small amount of sand or a stone is placed at the centre to make it hold an inverted cone shape. As the sun shines through the plastic, moisture in the sand is vaporised and condenses on the inside. As it builds up into droplets, these will run down the cone and drip off into the cup. The arrangement is illustrated in Figure 21.

The cone shape can be more readily obtained by cutting diagonally across a transparent plastic bag so that one of the closed corners can act as the apex. A 50 cm diameter hole about 40 cm deep is capable of providing 150-300 cm^3 (0.25-0.5 pint) of water per day.

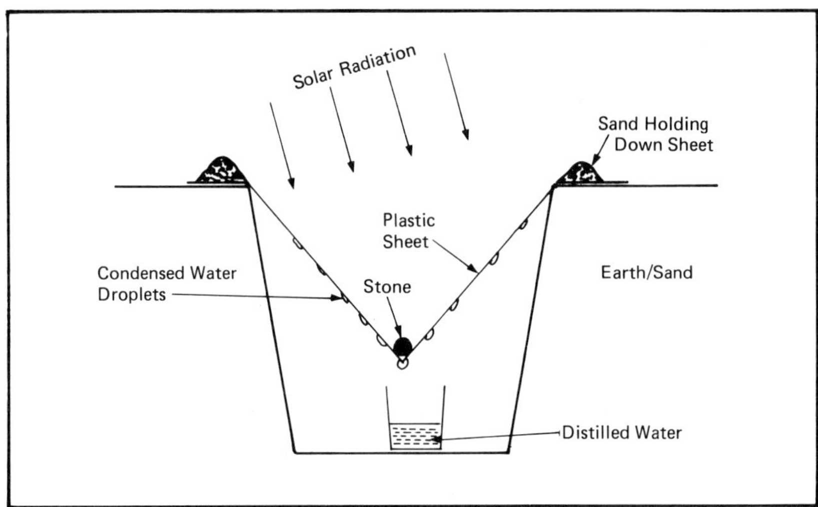

Figure 21.   Simple sand pit solar still

Another design which has aroused great interest is a vertical still (Figure 22) which possesses a very large effective surface area, as the brine, or impure water, is soaked up the length of a porous tube where it is readily vaporised by the sun's radiation. Modifications on the basic design enable it to be used to obtain pure water from the ground, or, by using a floating version, from sea water.

Another application of solar radiation that is not used a great deal in Great Britain but has potential for other countries is solar cooking; this does not mean that such systems cannot be used in this country, but simply that the solar conditions are not adequate enough to justify regular use of the sun's power for this purpose.

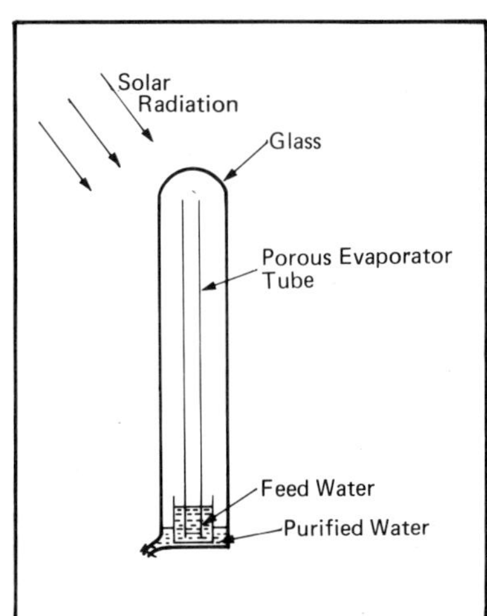

Figure 22.   Vertical solar still

As has already been explained, parabolic reflectors are used to concentrate solar radiation at one point or over a very small area, and it is with this type of concentrator that the most successful forms of solar cooking have been achieved. Thus parabolic reflectors

71

of 1.5-2 m diameter can heat 1 litre of water to boiling within 15 minutes when the water container is placed at the focal point of the reflector. In this type of cooker about 20 per cent of the radiation is actually used for heating to reach boiling, 45 per cent is lost by convection and 35 per cent is used in vaporising the water. For continuous heating over long periods the reflector needs to be repositioned every thirty minutes. For poor nations possessing hot climates this means there is an alternative to the use of animal dung and wood as a fuel, so long as cheap methods can be found for providing the parabolic reflectors needed for this purpose. Typical of the materials being tried at present are 48" diameter polystyrene parabolas with metal edges, covered with aluminised plastic film. These provide about 75 per cent reflection with a focal point 18" from the centre. Other designs have used anodised, spun aluminium. In all cases temperatures of 350-400°C can be obtained at midday in hot climates. Cooking by this method is limited to the periods of maximum sunshine on days which are both hot and clear, as the reflector only operates on direct sunlight and cannot use diffuse sunlight.

As an alternative to the parabolic reflector, it has been claimed that very high temperatures can be attained by use of a solar oven similar to the water heater previously mentioned on page 62. For this purpose an insulated oven with sides of about 25-30 cm (10-12 inches) is used (Figure 23). By having only one windowed side to the box and angled external reflector flaps in addition to the internal reflector surfaces, temperatures exceeding 200°C (390°F) have been reported. This type of oven is less dependent upon direct radiation, although it does benefit from being realigned with the sun every half an hour.

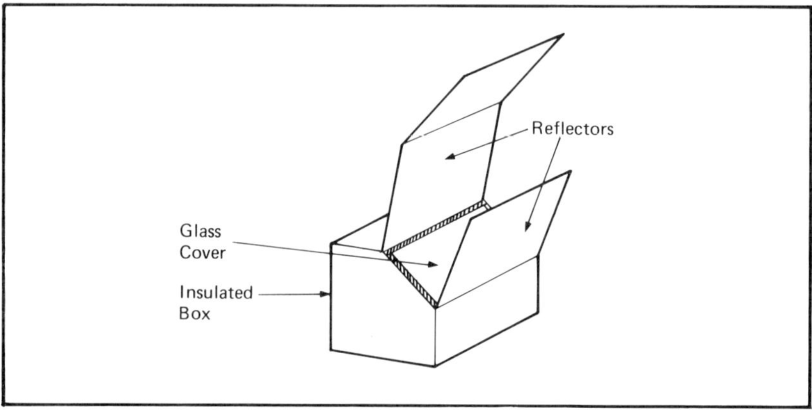

Figure 23.   Solar oven

There is nothing new in using solar radiation for cooking in this way. Tests on different designs of ovens have been carried out for many years in Africa, India, Mexico and other countries with some substantial success. There is, however, still scope for improvement in the design of the reflector materials and in developing units which are not only efficient but very cheap to manufacture on a large scale. Even if these objectives are attained, they need to be followed up by a widescale programme to teach potential users the advantages of solar ovens over established cooking methods. An obvious extension from the solar box oven is the use of solar radiation for drying. The exposure of wet objects, such as laundered sheets and clothes, to wind and sunshine for drying is common in most countries. It is also common to dry crops in this way. However, it has long been accepted that crops dried in the open air are subject to contamination from atmospheric pollution as well as being susceptible to being eaten by insects and rodents. At the same time forced drying by oil and gas heating can be expensive. Solar driers have been developed to overcome these problems in order to obtain uncontaminated, uniformly dried products. The simplest form of solar drier (Figure 24) consists of a box with a sloping glass lid and a shallow tray to hold the crops. This system is capable of drying root crops such as turnips, carrots and radishes in 6-10 hours, removing about 90 per cent of the moisture. The quantity of crops that can be dealt with depends upon the size of the solar drier, and some are big enough to be used for removing moisture from timber.

It should be obvious to the reader that although solar energy is not the absolute answer to the energy crisis, there are a number of advantages which are worthy of further study and greater application

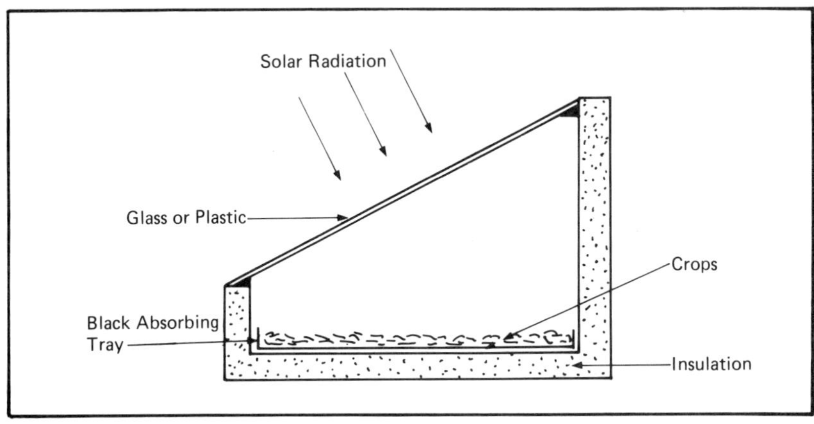

Figure 24.   Solar drier

within the home. As energy costs continue to increase, these advantages will become even greater, so long as the costs of the materials (including the raw materials) do not rise at the same rate as do oil, gas, coal and electricity.

4

LARGE SCALE
USE OF SOLAR ENERGY

Large-scale savings in the use of fossil fuels can only come about by the use of solar radiation if it is employed for industrial purposes. One of the most important questions which has not yet been satisfactorily answered is how this is to be done. Although there are individual instances of specialised applications, solar radiation is not yet very widely used in factories and offices. It is only in the past few years that any substantial consideration has been given to the wider aspects of industrial applications of solar radiation, although as recently as 1977 surveys on solar energy in Great Britain had shown that estimates of the benefits of solar energy to industry could not be made due to insufficient information being available. Fortunately this situation has been partially rectified as a result of the work of the U.K. Section of the International Solar Energy Society with its conferences and publications.

The development of the large solar distillation apparatus in Chile (see page 25) proved quite clearly that the use of solar radiation is not limited simply to domestic heating. It is also quite obvious that providing there is sufficient exposure area, it should be possible to provide many factories and offices with solar heating for their normal hot water supply for part of the year. The few special solar office buildings that have been built have shown that by combining the use of solar panels on roofs and walls with thermo-syphoning air and heat storage systems it is possible to provide large proportions of the heating requirements from the sun. The real problems come with seeking to use solar energy to obtain high industrial temperatures, for space heating of large working areas, for electrical power generation, and for mechanical power.

The use of solar radiation to produce very high temperatures involves using either large-radius convex lenses or large-curvature parabolic reflectors to focus the concentrated radiation upon a very small area. Large numbers of furnaces employing radiation concentrators have been built, but many use carbon arcs as the radiation source and only about a quarter have been built specifically as solar furnaces. The first of these was built in 1921 at Jena in Germany; since then many other solar furnaces have been constructed, many in the U.S.A. and several in Japan. A solar furnace built in Piran in Yugoslavia used a 1.5 m (5 ft) diameter searchlight reflector with a 65 cm (26 inch) focal length to produce temperatures exceeding $2000°C$ for the purpose of studies on the growth of crystals and the production of pure minerals. The solar radiation was directed onto the reflector by a series of mirrors forming a heliostat that could be controlled to follow the sun's path.

The most well known work in this field has been carried out under the direction of Dr. Felix Trombe. His first solar furnace, producing an output of 40 kilowatts, was built in Algeria when under French administration. This had the great advantage that it could operate for 80 per cent of the year because of the clear skies. This work was later transferred to France when Algeria was granted independence, and a new 50-kilowatt furnace was constructed in 1962 at Mont Louis in the Pyrenees close to the frontier with Spain. The furnace used a fixed parabolic collector built from 3,500 small mirrors to concentrate radiation reflected from a 100 m² moveable heliostat. The radiation was brought to focus on an area of 5 cm² producing 200 cal cm⁻² sec⁻¹ and enabling temperatures above $3000°C$ ($5400°F$) to be attained. Following the success of the Mont Louis solar furnace, an even larger 1000 kW furnace was also constructed as part of the French solar research programme at their Centre National de la Recherche Scientifique at Odeillo near to the Mont Louis site and was completed in 1970. The heliostat for this larger solar furnace (Plate XIV) consists of sixty-three mirrors on a hillside reflecting the radiation onto a 2000 m² parabolic concentrator made from 9,500 small mirrors and focused to provide a temperature of $4,000°C$ over an area of 30 cm² and $3,000°C$ over an area of 1250 cm². Active working at this site is restricted to about half the year, as it does not enjoy the very clear weather for which the original Algerian site was chosen.

Other, less powerful, solar furnaces developed in the U.S.A. have used large convex lenses with an area of 5 m² to attain temperatures of up to $3000°C$, and these have been equally useful for investigating the effect of high temperatures on substances in the absence of the gases and fumes associated with combustion methods of heating. Solar furnace studies have enabled accurate measurements of such things as the expansion coefficients and melting

76

Plate XIV.   Solar furnace at Odeillo, France

points of metals to be carried out. They are now fairly commonly used for such specialised material investigations and have been constructed in a number of countries where there are areas which enjoy long periods of direct solar radiation.

They do, however, all suffer from a number of disadvantages. The main one is that of cost, which is so great as to be prohibitive in most instances without either government finance or international collaboration. Secondly, as in all focused systems, only one side of the object is heated and although this is not of great importance on small samples it means that larger objects can suffer from temperature variations. To overcome this problem the samples under examination are rotated to obtain uniform heating from the very concentrated radiation.

It is quite obvious that the solar furnace is not likely to play a major role as an alternative form of industrial energy. Its main function will remain as a specialised procedure attaining elevated temperatures with some industrial applications but mainly of academic interest, and further developments will be chiefly in the building of small solar furnaces and the improvement of the efficiency of such devices.

Although solar furnaces are not likely to have a very wide application, it is possible to use the sun's power to provide mechanical power in a small way with a multiplicity of applications. The first attempt to use solar power to generate steam to drive engines was carried out by C. Günter in Austria more than one hundred years ago. To do this he used long parallel mirrors to reflect the sun's rays onto a long tubular water boiler. Similar attempts at steam generation for this purpose were carried out in France between 1860 and 1878 by A. Mouchot. At the Paris Exposition of 1878 another Frenchman, A. Pifre, used parabolic concentrating reflectors to produce steam for an engine that could drive a small printing press.

Since the mid-1960s a number of experimental solar steam boilers have been constructed. One developed in Genoa University, Italy, used mirrors to concentrate the radiation onto a 60 m (200 ft) long blackened steel tube through which water was passing and produced steam at 500°C (932°F). More recently solar power plants have been under consideration in Iran, where they are using parabolic concentrators to produce high-pressure steam at 300-500°C for power stations of 10-100 MW generating capacity.

In steam generation by solar power the major problem is to concentrate the radiation sufficiently to produce the high temperatures required. Even now, most designs are based upon that used by Shirman and Boyd for a solar-powered irrigation plant built in Egypt in 1913. This employed five parabolic concentrators which were over 60 m (200 ft) long and 4.25 m (14 ft) wide with the water/steam pipe fixed at the focal point along the length of each concentrator (Figure 25).

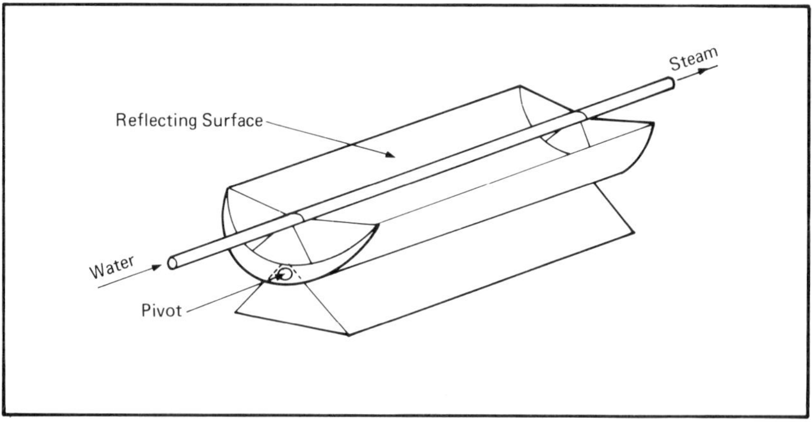

Figure 25.   Parabolic concentrator for steam generation

Similar designs employing a glass tube along the parabola axis have been used to heat air to 193°C (380°F) as the basis for gas turbines. Unlike steam, however, there is no substantial expansion in volume as occurs when water changes its physical form. So the potential for gas-turbine power from solar radiation must be rather limited.

Although such devices are unsuitable for generating electricity on the scale carried out by modern fossil-fuel power stations, the only limiting factor against small solar units for producing mechanical power in remote, hot areas is that of cost. Design improvements are overcoming this problem and greater use of steam power in the poorer countries of the world must be very near.

Another approach to the mechanical power problem has been the development of hot-air engines in which a sealed volume of gas is alternately cooled and heated — in this case the heating is derived from solar radiation. The idea of hot air engines is an old one, the first patent being given to Henry Wood as long ago as 1759, although credit for the first operating heat pump went to Sir George Cayley in 1807.

A great deal of research has been carried out on these engines because they are capable of providing a useful amount of mechanical power available for relatively simple purposes in the poorer countries. Again it is a matter of trying to produce a simple piece of mechanical equipment, which is robust and requires little maintenance, at very low cost. Many of these systems are developments and extensions upon what is known as 'Stirling's hot air engine', first described in 1816, in which heat is supplied to the closed air system which expands and contracts between two pistons. For solar energy purposes the source of the heat needs to be solar radiation concentrated and focused by a parabolic reflector. The operation of such a device, in a simplified form, can be followed in conjuction with Figure 26. The solar radiation, concentrated by a parabolic reflector, is directed onto the air above one piston, causing it to expand and drive the piston downwards, turning the flywheel. As the flywheel continues its rotary action it pulls down the second piston and drives up the first piston causing the expanded air to pass into the second cylinder where it loses some of its heat to a water jacket or through metal fins attached to the outer walls. Further rotation of the flywheel then pushes the second piston up and pulls the first piston down, driving the cooled air back into the heat absorption side of the engine, where the cycle can start over again.

The efficiences of many of the hot air engines have been quite

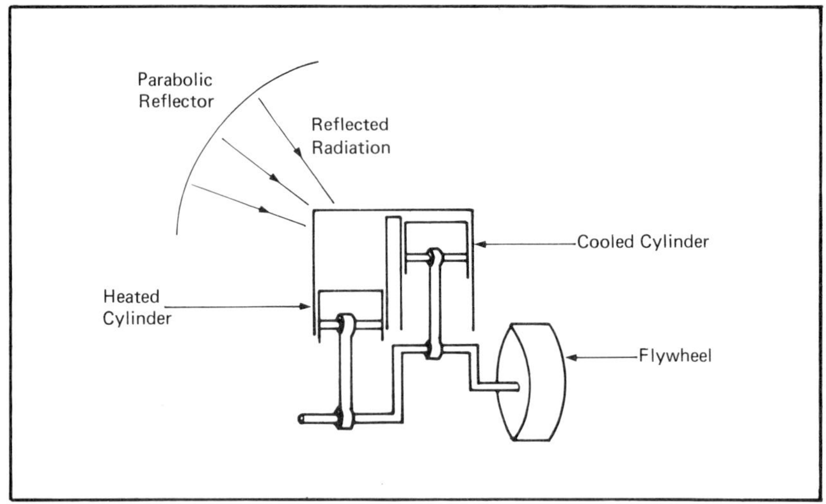

Figure 26.   Principle of Stirling's heat engine

low and research has been aimed at improving upon this by increasing the amount of heat transferred to the air in the expansion cylinder and the rate of cooling on the compression cylinder. Other variations on this basic design have included the incorporation of a quartz window above the 'hot' piston head so that the solar radiation can be focused direct into the air space instead of having to heat the cylinder head outside. Attempts have also been made to use gases other than air for the expansion-contraction cycle. Although no major breakthrough has been achieved in the use of these engines, they are capable of working with sources of light other than that from the sun, and even the focused radiation from lamps and flames can be sufficient to drive them.

Simple hot air engines based upon Stirling's principles are low-powered, less than 1 h.p., and relatively inefficient. They suffer from the disadvantage that they require a lot of heating before they begin to operate adequately, and this means that even in a hot climate a major part of the useful day may be wasted before the engine has warmed up. Because of this the development of more powerful pumps is based upon the use of special flat-plate solar collectors through which low boiling liquids are heated under pressure and then allowed to pass through an expansion chamber to drive a small turbine or piston engine. These appear to have greater potential and are being studied with considerable interest for the arid parts of the world to provide the pumping section of small irrigation stations.

80

The World Bank has financed a development programme to demonstrate the use of solar systems for irrigation in the poorer countries. Already field trials are in hand in Mali and Sudan employing both solar-thermal and photo-voltaic (electrical generating) systems. The consultants for this work, Sir William Halcrow and Partners, are also studying the lifetimes of components under laboratory-designed tests to ensure that any recommended device is not only cheap to produce, but also robust and capable of surviving rough treatment.

Of much greater general use throughout the world are heat pumps. These have the ability to take heat from a cold source and transfer it to a hot source, and can be considered to act in the same way as a refrigerator pump. The classical refrigeration system is based upon the circulation of a liquid, the refrigerant, which vaporises in the circulation plate within the refrigerator, taking up heat from the contents of the refrigerator. The refrigerant vapour is then passed through coils and a pump outside the refrigerator, during which process it is compressed back to a liquid and gives up the heat it had absorbed in becoming a vapour.

In the case of the heat pump, the system is used for heating purposes rather than cooling, and it can operate in such a way that in the cold months of the year it will take the colder air from outside the building and give up the energy as warm air inside the building. The whole idea of heat pumps was first proposed by Lord Kelvin more than one hundred years ago in an article entitled 'The Economy of Heating and Cooling Buildings by Means of Currents of Air'. Although he saw the potential of using the system for heating, it has been used almost entirely for cooling and air-conditioning, in which the heat is absorbed from inside the house and transferred to the air outside just as with refrigeration.

For a heat pump to operate satisfactorily, it requires a refrigerant possessing a boiling point below the temperature of the heat source — for house-heating purposes this means using a liquid such as Freon 12 which boils at -29.7°C. At the same time the vaporised refrigerant must be capable of being compressed by a low-powered compressor unit.

The first step which takes place in the heat pump is compression of the vapour at the external source temperature (this may actually be the external atmospheric temperature, the temperature of a lake of water, or heat from the ground). Compressing the vapour puts work into the thermodynamic system, and because it is under increased pressure it will boil at a higher than normal temperature designed to be above that required for the space heating to which it is to be applied. The compressed liquid is then passed through the expansion coils over which air or water is passing, and as this is below the boiling point of the refrigerant, the vapour condenses out, releasing its latent heat to the surroundings. The liquid refrigerant collects in a reservoir from which it is passed through a valve and allowed to expand with a simultaneous reduction in temperature taking place as it changes back into a gas. This low-temperature gas is then able to take heat from the surrounding heat source and the cycle commences again. A diagram of the essential parts of the heat pump is shown in Figure 27. The obvious question that people ask is how a system such as this which uses electricity for the pump can actually gain energy. In general practice the heat pump provides about three times the amount of energy required for its operation, and this is referred to as its coefficient of performance (COP). (Other COPs are possible, depending upon the temperature difference between the heat source and the heat sink.) It achieves this because the only use of energy is in the compression pump, whilst the gain in energy is the result of the heat acquired by the circulated refrigerant. The two factors are not necessarily directly related, as the amount of heat acquired by the refrigerant depends upon its physical properties, not just upon the efficiency of the compressor. It is, therefore, apparent that different refrigerants will give other COPs, and the selection of a suitable refrigerant depends very much upon the temperatures at which it is proposed to collect and deliver the heat. Success depends upon having a refrigerant which has a high latent heat of vaporisation and takes a

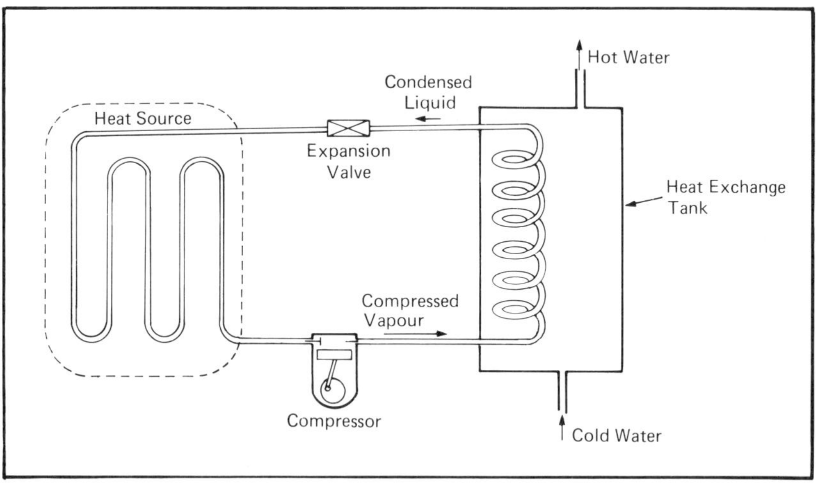

Figure 27.   Heat pump circuit

82

large amount of heat from the source in the process of turning from a liquid to a gas. However the COP is highest when the temperature difference between the source and the sink is smallest.

One very important feature of heat pumps is that they need an adequate source of moderate temperature heat from which heat can be taken and passed to the heat sink. Three sources are normally used for this: heat from the ground, heat from a river or lake, or heat from the atmosphere. In all three instances a large surface area of piping is necessary, and for most domestic hot water supply systems the obvious source to use is the earth, with pipes buried several feet below the ground. Under normal conditions that will ensure a fairly constant supply of heat, replenished by further heat from the atmosphere.

As the heat source end of the heat pump is, in effect, the cold end of a refrigerator, it is also possible to utilise that end for cooling while the heat is being provided at the other end. Such systems taking their heat from the atmosphere have been used for cooling apple stores and larders.

For more substantial applications of heat pumps on an industrial scale it is possible to take the residual heat from a water-cooling tower and by running the water over the heat-extracting coils of the heat pump system utilise it for providing heat to the factory. This is a classic method of recycling waste heat.

From a domestic point of view the major disadvantage of the heat pump system is its expense, and this even applies it if is compared with the cost of installing solar panels. The additional expense arises from the need to use a compressor instead of a small circulating pump, and a much larger surface area of heat-exchanger tubing. A typical domestic heat pump requires 100-150 m (approx. 300-450 feet) of 19 mm (0.75 inch) copper tubing buried about 1 m (approx. 3 feet) in the ground, or at the bottom of a pond. It should be capable of taking heat from the source at about 2-5°C (35-40°F) and supplying heat at 49-60°C (120-140°F). These factors mean that the pay-back period for a heat pump is far greater than that of most other forms of fuel economy, although it is only fair to say that many heat pumps have operated over very long periods with very little maintenance and few running problems.

Another expense feature of the heat pump which has to be borne in mind is that it does use a fair amount of electricity, and the benefits of the heat pump will depend greatly upon the relative cost of electricity compared with other forms of energy. It is important to remember, too, that even with a COP of 3 the electricity is giving less than an 80 per cent return on its primary

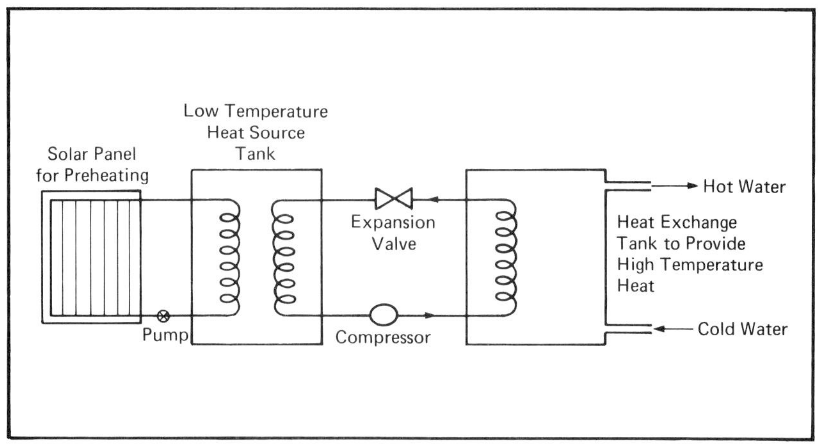

Figure 28.   Combined heat pump-solar panel circuit

energy source, as it is only generated with 28 per cent efficiency. There is, however, one great gain with heat pumps, and that is that they can be used to provide space heating, either by feeding hot water to radiators or by heat exchange with a circulating air system. In these cases the temperature of the heated supply is lower than that normally obtained from conventional gas or oil-fired boilers, and for adequate heating it is necessary to use large-panelled radiators. More efficient space heating can be obtained if solar panels are used to provide a low-temperature heat reservoir as the source heat for the heat pump. This enables the heat pump to produce an even higher temperature for the water radiators or the ducted air. This type of arrangement is shown diagrammatically in Figure 28.

In the U.S.A. there is increasing interest in heat pumps because of their potential both for space heating and for air conditioning. It is the air-conditioning aspect which has traditionally received the greatest study over the years, as that is cheaper to operate than is the space heating. However, with the increased cost of oil and the improvement in the design and operation of dual systems, the all-purpose heating and cooling circuit is becoming more common. Most of these tend to be based upon taking the heat from and releasing it to the air, as they are used in urban areas where it is impractical in most cases to put large lengths of tubing in the earth or into the river. One problem that can arise with these is that in the winter the external heat-source coils cool and can become frosted over, and the layer of ice crystals acts as an insulation which reduces the heat transfer between the external

84

air and the refrigerant. When this occurs it is necessary to reverse the cycle in order to heat the coils to defrost them.

This defrosting problem does not occur with coils immersed in the ground, although it has been reported that a layer of snow on the ground also has an insulating effect when present for an extended period; is imposes a greater strain on the compressor, as the earth is then becoming progressively colder instead of being able to replace its heat from the atmosphere.

The further installation of heat pumps will depend very much upon the rate at which energy prices are increased. Most do-it-yourself exponents have tended to steer clear of this type of installation as, unlike solar panels, it does not come in easily assembled do-it-yourself kits. But if the demand for heat pumps increases there will be a corresponding increase in people seeking to install their own systems. However, what is needed at present are more precise figures on the cost of installation and the true pay-back period after allowing for depreciation, maintenance and interest on the value of the capital investment. Such figures, when available, tend to be highly conflicting.

For industrial space heating there is great potential for heat pumps on factory sites where much surplus heat is wasted instead of being recycled. The wider use of solar energy in all its forms in the working environment is, however, still dependent upon companies making a careful assessment of the cost of the energy they use and the amount which is wasted because it has traditionally been cheap.

5

ELECTRICITY FROM SOLAR ENERGY

For many years people have sought to overcome the energy problem by using solar radiation to produce electricity. If that could be easily and cheaply achieved, then electricity would become the prime source of power for motor vehicles, trains, heating, lighting and industry, and fossil fuels could be reserved for the more important aspects of converting to necessary chemicals and industrial materials. So far the magic answer has eluded the researchers and there is no cheap way to generate electricity from the sun — it can be done, but only at a price that places it way beyond established electricity generating costs.

Many people are familiar with what is commonly called the 'Seebeck effect', named after its discoverer in 1821. This is the basis of the thermocouple, in which two dissimilar metals are joined to produce two junctions which are maintained at different temperatures. The difference in temperature causes a small current to flow that is characteristic of the metals forming the thermocouple and proportional to the temperature difference between the two junctions. This system is actually used for the measurement of temperature differences, which can be accurately calculated from the magnitude of the current flowing. Such thermocouples are made from alloys of chromium, aluminium, copper and nickel but their electric output is very small, between 20-70 μV for every degree Celsius temperature difference. Obviously a small current can be obtained if one junction is exposed to solar radiation and the other is kept at a lower temperature.

The application of metal thermocouples for electricity generation is obviously rather limited, but the development of semiconductors has changed this, as thermocouples made from these materials are capable of producing potentials of between 100 and 1000 μV for every degree difference between the junctions. These are employed

86

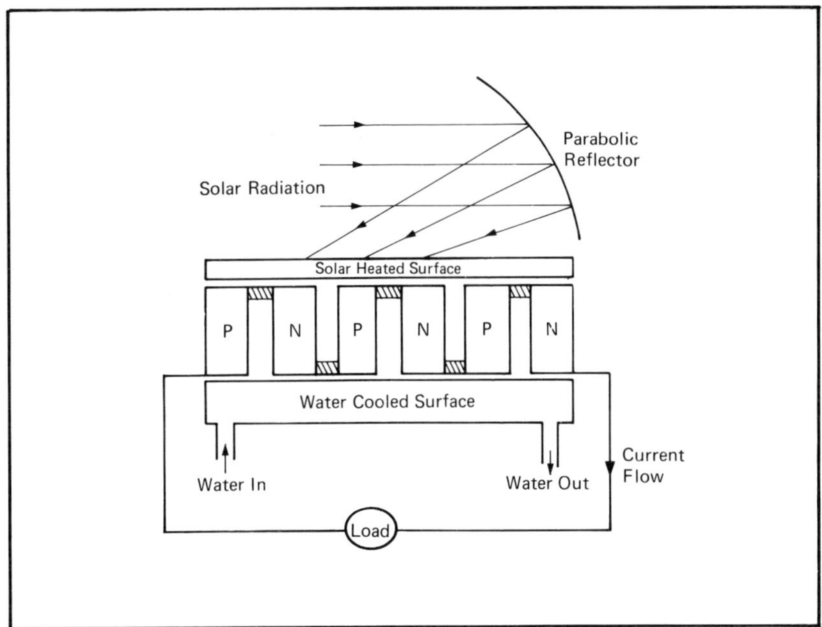

Figure 29.   P/N Thermocouple

in what are known as thermoelectric generators and use alternate blocks of N-semiconductors (those possessing free moving electrons) and P-semiconductors (possessing a surplus of positive charges) placed in series (Figure 29). Both sets of surplus charges can move, and do so at different rates due to the temperature difference between the ends of the elements, and it is this movement of charges which produces the voltage in the circuit. The efficiency of the system is dependent upon the temperature difference between the ends of the elements and the characteristics of the semiconductors employed. As a result of these developments the efficiency of thermocouples has been extended beyond the level of 1-3 per cent obtainable twenty years ago to about 7 per cent or even higher at this time, and research is still being carried out with the expectation that this will be further improved as new semiconductors are developed. By using the present semiconductor thermocouples linked together it is possible to generate fairly high voltages. Systems have been successfully operated with temperature differences of 500-1000°C and with sufficient elements to produce potential differences of several volts. The solar radiation used to obtain the elevated temperatures is focused onto the absorbent-coated upper surface of the generator using a curved-surface collector. As with most electrical generating systems these are normally operated in conjunction with battery storage arrangements, so that any surplus power obtained due to the solar radiation is stored for use when the solar radiation is no longer available.

From a purely practical point of view, the preferred method for obtaining electricity direct from the sun's radiation is by the use of silicon solar cells. These have the great advantage over thermo-couples that they do not require a temperature difference between a cold and hot area. An additional advantage is that they will function even in the absence of direct sunlight, so that an output can be expected during the hours of daylight even on a day when the sky is completely overcast. Enormous advances have been made in the development of solar cells, mainly due to the research that was carried out for the United States space programme. This work was done in order to develop systems which could provide the maximum electricity for the space crafts with the minimum of weight. Even so, vast numbers of solar cells have been needed to obtain the required currents and voltages. Mariner IV, for instance, used 28,224 silicon cells built into four vanes. The solar collector areas of the space craft look a bit like the arms of a windmill and unfold like the pages of a book after the space craft is in orbit. The lifetime of such collectors is almost indefinite and the use of them has enabled the space probes to be sent on extended journeys of many years into outer space.

From an earth-bound point of view solar cells are of considerable interest and will be of growing importance. Historically they have been known as photovoltaic cells since their discovery over one hundred years ago, when it was found that direct sunlight falling on wafers of selenium resulted in 0.6 per cent conversion of the radiant energy into electricity. Even as recently as 1953 it was considered that the limit of conversion was about 1.0 per cent, and the great advances beyond this have all been made in the last twenty-five years as a result of the introduction of semiconductors. The modern selenium cells have efficiencies as high as 25 per cent, whilst those made from gallium arsenide operate with 10-15 per cent efficiencies.

Photovoltaic cells are sometimes called barrier layer cells as their operation is dependent upon a junction (or barrier) between two layers of silicon. The positive side of the barrier layer cell is a P-type semiconductor (with a slight surplus of positive charges) formed from adding a very small quantity of boron to the silicon. The negative side of the cell is an N-type semiconductor (with a surplus of negative charges) obtained by adding a small quantity of

Plate XV.   Solar cell array

phosphorus to the silicon. The surface of the cell is fitted with metal finger contacts and is given a special coating that will assist it to absorb solar radiation. When radiation strikes the surface the energy is absorbed and leads to electrons crossing the P-N junction creating a potential difference in the circuit.

A single cell measuring about 4 cm^2 is capable of producing 0.4-0.5 V at 60 mA, whereas the early selenium cells produced 500 μV at 600 μA. The real question at present is not 'can solar energy be used to produce electricity directly', but can it be done cheaply. A single selenium cell costs between £1 and £1.50 ($2-$3), and to obtain any reasonable output of electricity they are purchased in modules containing a number of cells linked in series (Plate XV). A typical module might contain up to forty or fifty cells protected by a transparent cover and used to provide a charge of about 12-15 Volts d.c. Such a system can be used for many purposes, and any surplus current generated may be stored in a long-life lead battery for use when the cells are not able to produce sufficient current for normal use. Solar conversion efficiences can be improved up to the 40-45 per cent level by building what have been called 'Tandem cells'. These consist of both P- and N-type layers of two or three different materials (Figure 30), and the increase in efficiency comes about because the materials are chosen to respond to different parts of the solar spectrum. Further investi-

89

gations into this form of combination cell are being carried out.

Apart from the applications in providing electricity for space craft, solar cells are being used extensively for supplying electricity to remote sites where the laying of electricity cables over long distances would be prohibitive. In most instances the solar collectors are backed up by lead-acid or nickel-cadmium batteries for storage. One of the world's largest suppliers of solar cells, Lucas Electrical Ltd. of Great Britain, has provided solar generators for places as far apart as St. John's in Newfoundland, Cape Rodney in New Zealand and Qatar in the Arabian Gulf (Plate XVI). Fairly typical of the use made of these solar collectors is the provision of a radio-repeater station for the coastguard on the Isle of Islay in Scotland. The self-contained power system installed had to be totally reliable and able to withstand severe weather conditions. For this purpose the necessary supply was obtained from a 12 ft^2 (1.1 m^2) south-facing solar cell array angled at 55° to the horizontal capable of providing 1 amp at 24 Volts and charging a series of lead acid batteries. The work-load on this system operating a VHF radio is a continuous 2.5 Watts for the receiver plus a short term higher load of 30 W for about three-quarters of an hour every day. Of particular interest in the poorer parts of the world has been the use of silicon solar cells to collect energy to power 32 V television sets with 61 cm (24 inch) screens in Africa since 1968. The sets operate at 32 W using the energy stored in 16 x 2V batteries providing 1 A per hour.

From these examples, it is obvious that the potential for solar cell electricity generation is vast, with applications varying from the operation of water pumps for irrigation, to security devices

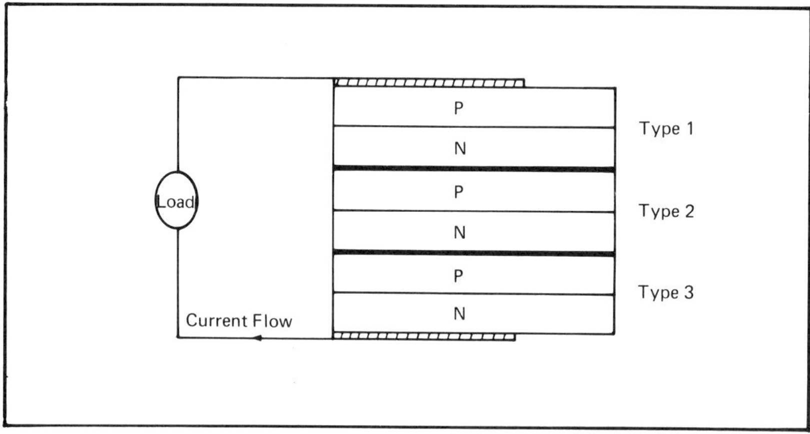

Figure 30.   Tandem cell

Plate XVI.   Solar cells in use

and navigational aids for unmanned lighthouses and buoys. Maximum output is obtained, as with all solar devices, by the solar cell array being orientated towards the equator. As most systems are intended to provide current around the year the optimum angle of inclination to the horizontal should be identical with the latitude of the site. It would be wrong to assume that such systems are only suitable for government installations. They can not only be used to provide enough current for domestic lighting but also to run low-power electrical equipment such as electric razors and even refrigerators.

Various efforts have been made to increase the output obtainable from solar cells. Typical of the approaches made have been those in which the cell is situated at the base of a small parabolic reflector concentrator. By this means the output of the cell can be increased by a factor of one hundred. Another approach has combined the solar cells with an ordinary flat-plate collector, as has been done in the very advanced solar house built at the University of Delaware, U.S.A. The cells employed here are cadmium sulphide photovoltaic cells sealed behind double-glazed panels using plexiglas. Of the total radiation reaching the cells 5-10 per cent is converted to electricity and the remainder used to heat the air, which is first ducted behind the cells to keep them cool. As with other solar cells the efficiency of the cadmium sulphide cells decreases as the cell temperature increases, so that the cooling process is of considerable importance in really hot weather.

One feature of solar cells that should not be ignored is that they can be used to power pumps, and can be fully integrated with any solar-panel hot water system so long as a back-up battery system is employed for those periods when there is not sufficient solar power to operate the pump. This type of arrangement has been used with some of the roof-mounted solar panels at the National Centre for Alternative Technology in Machynlleth, Wales (Plate V). For this purpose solar cells are used to provide about 12V to power a 30W circulating pump through a storage battery which can operate at 6kWh.

It should be quite evident that three major problems arise from the use of solar cells. The first of these is that the quantity of electricity produced per cell is very small and for any reasonable output large areas of cells are frequently required. Most of the permanent arrangements for isolated uses in telephone boxes, marine installations and radio receivers are designed to operate with low voltages and amperages and are unsuitable for the average household employing either 120 or 240 volts.

92

The second problem is one of cost. Although the basic materials in solar cells are cheap, their manufacture in a usable form is extremely expensive. As a result solar electricity is 50-100 times more expensive than that obtained by traditional power generation. One of the major fields of research is the endeavour to reduce this cost by simplifying the manufacturing process or by finding substitute materials for the silicon. Although other materials such as gallium arsenide and cadmium sulphide are not as efficient as silicon, there is every likelihood that they may be proportionally cheaper to produce.

The third problem is that the cells cannot operate when the weather is bad, and some form of storage must be provided in order that the electricity supply can be maintained during these periods. Traditionally lead/acid or nickel/cadmium batteries have been used for this purpose, but these are also quite expensive and research is now directed to looking for alternative battery systems. The energy obtained from solar cells can also be stored by converting it to other forms of energy and then reconverting that energy back to electricity if necessary. This may sound a rather long way round, but it can be economically justified. It is an approach that is already used in some places where surplus electricity from power stations is used to pump water up to a reservoir (the electrical energy is converted to potential energy due to the height of the water) and the water is allowed to flow down and turn generators when peak demand for electricity arises (the potential energy is converted to mechanical energy which is then converted to electrical energy). In this process there is an enormous loss in the overall energy equation and the amount of electrical power obtained at the end is only a fraction of that started with. However, even with a small solar panel it is possible to power a small water pump which can be used to fill a reservoir that can in turn be used to drive a low energy generator.

Another method of storing electrical energy as mechanical energy is by the use of a flywheel. In this approach the electrical power is used to drive a motor which turns an axle on which is fitted a large flywheel. So long as the electrical power is sufficient to overcome the friction of the mechanical system the flywheel will progressively gain speed and hence kinetic energy. When the solar power drops, the rotating flywheel can then be used to drive a generator to produce electricity. As it does so it will, of course, slowly lose speed until there is additional solar power available to keep it moving. Not only do flywheels offer prospects for the efficient large-scale storage of peak-load electricity, but they may also be used in compact units for powering electric vehicles. In the former case 4.5 m (15 ft) diameter flywheels weighing 200 tonnes and capable of storing 20,000 kWhμ of energy by rotating at 3,500 rev min^{-1} are designed. The limits of size and speed are set by the strength of the materials, as the amount of energy that can be stored depends upon the mass of the rim and the rate of rotation. For motor vehicles it has been calculated that a 100 kg (220 lb) flywheel could store 30 kWhμ of power to drive a vehicle for 320 km (200 miles) at nearly 100 km hμ^{-1} (60 m.p.h.).

At least for the time being it looks as if the storing of electrical power will be carried out mainly in batteries, and research in this area will be devoted to new materials and reductions in battery weight and size, as well as to storage capacity. Enormous advances have already been made in the field of solar electricity generation and there is every possibility that with the increasing concern over the world's energy needs many of the present problems will be well on the way to being solved in the next few years.

Although such things as solar-powered digital watches may be novelties at this time, there will be an increasing application of solar cells for everyday purposes over the next few years. Along with the microprocessor and silicon chips, the solar cell will be used to provide power for rechargeable batteries in radio sets, tape recorders and other relatively low electrical energy users. Already it has been shown that with sufficient solar cells it is possible to power a small motor vehicle and a motor boat. Although it is unlikely that solar energy will ever generate electricity on the level of the large electricity power stations, it could become a useful supplementary source for the domestic and small industrial user.

6

THE FUTURE ROLE
OF SOLAR ENERGY

Whenever an attempt is made to anticipate what will take place in the future there is a tendency for people to assume that nothing of value has been achieved that can be applied to the present. But with solar energy it is not just a case of 'jam tomorrow'. It is very important to appreciate that in parts of the world solar energy is already contributing very usefully to energy needs, particularly in countries such as Israel which are virtually starved of sources of fossil fuels. Even without further scientific and engineering advances being made, there is no doubt that applications of solar energy for heating, cooling, drying and purification of water will be greatly increased in the immediate future. Even the oil-rich nations can foresee the days when their wells will run dry and are looking at what solar energy has to offer them in their sun-blessed, as well as oil-blessed, lands.

Greater use of established solar collectors for domestic heating will depend upon economics and incentives. Few people are prepared to spend large sums of money to install systems which will not repay their costs under 10-20 years, as most of them will either move or die during that time. Some countries, such as the U.S.A., have overcome this problem by providing grants or tax concessions to encourage the installation of solar collectors. In Great Britain there is a disincentive, as the installation of solar collectors increases the rateable value of the property and hence the householder has to pay an increase in the annual rates, just as he does if he puts in double glazing. The future for solar water heaters in Great Britain will continue to be relatively poor unless the manufacturers can manage to reduce the installation cost, in the absence of any government steps that will encourage more people to invest in such devices.

In the past another disincentive to using solar energy has been the possibility of spending a lot of money on an inferior system installed by one of the 'cowboy' companies that jumped into the sunlight when the energy crisis gained momentum. Fortunately many of these have gone out of business very quickly and the industry, through the Solar Trade Association, has taken its own steps to safeguard the purchaser.

At the same time a British Standard for 'Solar Heating Systems for Domestic Hot Water' has been drawn up. This covers the type and performance of all the materials used in solar panels in addition to dealing with the planning requirements to satisfy the Building Acts and the Water Acts. It is a useful guide for the prospective purchaser of solar panels as well as for the manufacturer and installer.

The future use of solar energy will fall into three main categories. There will first of all be the low-cost use of solar radiation in terms of the established technology for heating swimming pools, domestic hot water, water purification and irrigation, solar cooking systems, with further developments in terms of coupled solar collectors and heat pumps to provide central heating and air conditioning. Improvements that are taking place and will take place in this sphere are those concerned with the development of better materials requiring less attention and fewer replacements. This realm of application has already been shown to be perfectly viable; it is now up to private companies with government encouragement to make the applications more acceptable to the general public.

The medium-cost applications of solar energy lie in the realms of electricity generation and the biological conversion of the sun's energy. The widespread use of solar cells, other than in novelty features or for isolated sites, will not come about until there is a substantial reduction in the price of the solar cells. Some reduction can, in any case, be expected as a result of larger-scale production, and their use will become more economically justified as other energy sources increase in price. There is, even now, no reason why greater use of solar cells backed up by battery storage should not be made in places such as those used seasonally for camping and caravanning.

The largest world usage of solar energy is in the photosynthesis carried out by plants, a process of very low efficiency but of considerable importance to man and to this planet. If trees can be made to grow more rapidly, then the energy problem can be solved and timber may once again replace the plastics obtained from petrochemicals. This is the thinking behind the research being carried out into biochemical conversion systems — the so-called 'biomass' studies. At the same time work related to this includes
96

biological engineering to develop new strains of plants that will grow faster, larger and stronger to provide food and fuel for the ever-increasing world population.

At present only about one-fiftieth of one per cent of the solar radiation reaching the Earth is usefully converted by agriculture. In terms of weight of growth of plants, photosynthesis can produce up to $9 \, kg \, m^{-2} year^{-1}$ ($1.8 \, lb \, ft^{-2} year^{-1}$) of biomass for rapidly growing substances such as sugar cane, although for grass crops like rice and corn the amount of biomass is about one-tenth of this level.

Part of the work on biomass includes studies on highly photo-active algae which carry out photosynthesis with an efficiency of 5-10 per cent. The algae can be used as an animal feedstuff, burnt to generate electricity, or fermented to produce methane, and have the advantage that they can be grown on liquid and solid domestic waste.

In the year 1977/8 the U.S.A. had a research budget of $22 million for biomass alone, with particular emphasis being given to various procedures for producing large quantities of alcohol. In France it has been calculated that 11-14 per cent of her total energy needs can be provided from biomass, whilst several countries in addition to Brazil are adding increasing amounts of biomass alcohol to petrol for motor vehicles.

It should, however, be emphasised that obtaining ethanol from plants to use as a fuel or chemical starting material is not at present always economically viable. In some cases, depending upon the plant source of the alcohol, the distillation to purify the alcohol may consume more energy than is available from the product. It is because of this that the best plants for ethanol production are those with high sugar or other carbohydrate levels. For this reason some of the current solar energy research is concerned with finding rapidly growing high-level sugar-producing plants to improve the economics of alcohol production.

Although 'biomass' studies are costing a great deal of research money, it is possible for every person to play a small part in using established knowledge to use this form of energy conversion. The most obvious way is to plant trees, not just because they look nice but as a crop, to provide wood both in the near and distant future. Vast areas of the world are barren and in some cases deserts have developed because no replanting was carried out when trees were cut down to build ships, houses, to make paper and for fuel. There are many advantages which go with tree planting, not least of which is the overall improvement to the environment. The many empty areas around motorway junctions, on housing estates and in con-creted shopping centres can all be used to provide the many useful and important materials that can be obtained from wood and that will be needed to replace the petrochemicals in twenty or thirty years time.

One 'biomass' development that may well occur in the relatively near future is the home production of alcoholic fuel by fermentation and solar distillation in order to power the future version of the motor car. Already many people make their own beer and wine at home; with the development of super moulds and yeasts it will easily be possible to ferment most of the vegetable and animal wastes at present thrown in the waste bin to produce enough alcohol solution in each home to run a car for an average mileage daily. Again, this is not a matter of 'Is it possible?', but rather one of 'Will any government be prepared to allow it to take place?' One thing is quite certain: the many millions of people with motor cars will not willingly forego the convenience of being able to travel in comfort as and when they wish. It has already been shown that they are prepared to pay a very heavy price to stay mobile, and there is no reason to believe that anything other than a complete lack of liquid fuel will take them off the road. The only alternatives to the introduction of alternative liquid fuels will be either improved electric-battery-driven cars or solar-powered cars using more power-ful solar cells with battery power storage. At present biomass liquid fuel is about double the cost of petroleum. But as the oil nations increase the cost of crude oil so biomass-produced chemicals are becoming more economically viable not only for fuel purposes but also as raw materials for the chemical industry. In thirty years' time we may well see the petrochemical industry giving way to a sugar-based chemical industry.

High-cost applications are those which can only be undertaken by the most advanced and wealthy nations in the world because of the enormous investment that will be needed for the initial research before any financial return can be expected from the investment. Typical of this is the construction of solar furnaces suitable for the production of high-pressure steam for electricity generation. The present stage of designing has reached the level of employing a 400,000 m^2 heliostat to concentrate the sun's rays on a boiler to produce enough steam at 500°C to generate 10 MW of electricity. This is, of course, small compared with the size of the 2,000 MW nuclear power stations now being constructed. But it is quite clear that it is already possible to provide small communities with electricity generated from steam heated from solar energy, and this should give some hope to the poorer nations which are con-centrated in the hotter parts of the world. However, at present not even the most optimistic researcher believes that power stations of this type will ever compete with traditional forms of obtaining electricity.

The real visionaries of solar radiation are looking at the possibility of obtaining electricity from solar radiation collected by large satellites orbiting the earth. It is envisaged that these solar space stations will use enormous panels of solar cells to convert the sun's radiation (which is much stronger outside the earth's atmosphere) into electricity, which would be further converted within the satellite to produce microwaves suitable for beaming to earth (Figure 31). This idea was first suggested by Dr. Peter Glaser in 1969, and since then the concept has been taken and extended, although Dr. Glaser considered that the development of such power stations was several decades away.

To operate successfully solar satellites have to be in stationary orbit, in order to transmit their radiation to a fixed area on the earth's surface. For this purpose they need to be placed very accurately about 36,000 km (22,300 miles) above the earth. Such precise placing is no great problem, but the magnitude of the undertaking is vast. It has been estimated that the solar satellites would need to be 11-18 km across in order to produce 5,000-10,000 MW of power. The energy would be beamed back to earth via a curved transmitter of 1 km diameter and would be received on Earth at a receiving station covering an area of 50 km^2 where the microwave energy would be converted back to electrical power. For a constant supply of energy at any point on the Earth's

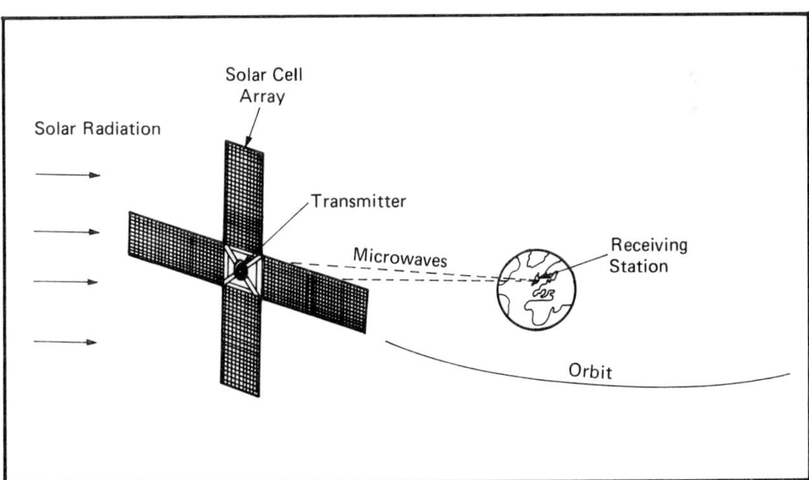

Figure 31.   Solar space station concept

surface a minimum of two solar satellites would be required, spaced 21° apart in order that one is always receiving solar power if the other is covered by the Earth's shadow. For such solar power stations to be feasible it has been estimated that component costs need to be reduced by a factor of 100 and orbit projection costs by a factor of 10.

It is worth pointing out that to provide the total electrical energy requirements of the whole world by the end of this century something of the order of 300 to 500 of these solar satellites would be required. Although the technology to achieve this exists, it is questionable if the cost of time, money, raw materials and energy for production and rocket propulsion justifies taking the concept beyond the limits of being a paper exercise. If the large sums of money are available they could be better spent on improving the established solar energy techniques on this planet. The great danger of vast projects such as solar satellites is that if applied in practice they make even more people dependent upon a centralised production of energy — that can always be switched off at the whim of a handful of operators or the action of a foreign power. The great value of solar power in all its earth-bound applications is that it makes the user more independent of the energy conglomerates and increases his self-reliance. For these reasons alone the use of solar power deserves much greater encouragement than it has hitherto.

The big question really is 'What has solar energy to offer for present generations as well as for those yet unborn?' Part of that answer depends upon how much money and effort is put into solar research over the next ten years and how rapidly the oil-producing countries by raising the price of their crude oil price themselves out of the market. One thing is already certain — there will be an increase in the number of solar collectors installed on the roofs of houses that have already been built. It is also clear that it is much cheaper for the purchaser if such installations are made at the time that the house is constructed, and following the experiences of a number of the larger house-building companies, solar collectors may become an optional extra for purchasers of new homes. There will also be a greater use of passive heat collection systems based upon the Trombe wall as used at Higher Bebington. At the same time architects and town planners are considering the building of what have been called 'ecologically sensitive housing estates' incorporating modern ideas of solar heating, energy conservation and integration of design.

For general purposes, the two areas with the greatest potential for real energy saving are the use of heat pumps and solar cells. Most of the development work for the first of these has already been done, and we have reached a stage at which the manufacturers need to be promoting a home installation package which can be adapted to any standard house which has an adequate garden in which the pipes can be buried to obtain the heat for transferring to the house. For solar cells the major problem to be overcome is that of cost and electrical storage. Substantial reductions in price have already been achieved in just a few years, and another ten or twenty years will almost certainly see a major transformation in this area of technology. Certainly these will be of the greatest value to the underdeveloped countries, as one of their greatest needs is for cheap, reliable, easily maintained, low-power systems to pump water and to provide lighting.

Although it cannot be claimed that solar energy is the answer to the world's energy problems, there is no doubt that it does have a major role to play. With even the major oil-producing nations looking to the sun's power to provide some of their own energy needs once the oil wells run dry, there is great potential for manufacturers and exporters. At the same time it is not only the hot countries that stand to benefit in this way. It is fair to say that anywhere that plants grow and carry out photosynthesis there is scope to utilise solar energy in other ways. The industrialised world is desperate for supplies of energy that will make them independent of the oil producers, and they cannot afford to continue to turn a blind eye to the endless supply of free energy from the sun. Even a 10 per cent saving on energy imports is worth having.

For the ordinary person who has electricity, gas and petrol bills to pay, the whole thing is even more fundamental in that it must be shown that there is a financial advantage in using solar installations. At present there is no simple answer. The economics of the matter are moving in favour of domestic solar installation all the time. Whereas a few years ago it was only economically justified to install solar collectors in places such as Florida, Texas, Japan, Israel and Australia, the latest oil and gas price increases have made it marginally worthwhile for even the coldest parts of the world. However, the calculation is a relatively simple one for anyone trying to decide whether or not to go ahead with flat-plate solar collectors for domestic hot water. At the very best solar collectors will provide about 50 per cent of the heating for the hot water, which in the average house is just over 20 per cent of the total heating bill. This means that the annual saving on the total energy bill (electricity plus gas plus solid fuel plus oil) will be about

10 per cent. In a typical centrally heated house this saving will be of the order of £40-£50 per annum. This saving must then be offset against the cost of any installation (including any interest charges and cost of maintenance) divided by the number of years of the anticipated life of the installation.

This will, of course, only give the rate of return based upon current energy costs. If some allowance is made for anticipated increases in fuel costs it is not difficult to see that a normal domestic installation should pay for itself within 10-15 years.

Although the use of flat-plate collectors is the most likely form of solar radiation that will affect the average person, a much wider result of the use of the sun's energy will be seen over a number of years in changes in building design. Architects and builders are already designing houses, schools, offices and factories with much greater regard for heat gains and losses. Not only is consideration being given to the insulation value of the actual materials used in the construction but also to the general circulation of air within the building and to the orientation of the building relative to the sun in order to achieve the maximum natural heating from solar radiation. The general approach for the future will be to orientate buildings with the largest exposed wall facing the sun. In some instances solar ponds, to absorb the sun's energy, will be incorporated onto flat roofs, and in other cases heat losses will be reduced by building up earth banks around the house.

This latter concept is no figment of the imagination. Several solar houses have already been constructed in the U.S.A. with earth banks against the outside walls and with gardens covering the whole roof. An attempt at modern cave living is now being tried by an architect in England. At the village of Holme in the Peak District National Park, Arthur Quamby has constructed a complete subterranean home, aptly called 'Underhill'. It has been cut into a rock face and has earth banks around exposed walls. A controlled climate has been created, assisted by a large interior swimming pool and massive concrete construction. Although not quite the type of thing that can be produced on a large scale it does show how it is possible to build without intruding greatly on a natural landscape. Now another land owner, in Amersham, has applied for planning permission to construct a complete concrete-walled house below ground level, with only a couple of greenhouses and plastic dome roof lights being visible on the surface.

Although dwellings such as these are exceptional, they do give an indication of how modern building materials and designs are developing and may be applied in the not too distant future.

It does not mean that future generations will lead subterranean, totally self-sufficient lives, but that greater consideration will be given to energy conservation, energy storage and the use of natural systems.

Buildings with large areas of single-glazed windows will no longer be desirable in an energy-conscious community where the need will be to obtain the maximum heating for the minimum cost. Future buildings will be designed to use heat exchangers and heat pumps to extract surplus heat from discarded waste bath and washing water in order to use it for preheating other water in the domestic water supply. This will be particularly applicable in large office blocks and factories; at the same time surplus heat from power stations and factories will be used for heating local housing areas.

The contribution that solar energy can make will vary greatly from one part of the world to another. But whereas in the past it has normally been taken for granted just in promoting photo-synthesis in plants and for toasting white bodies a delicate shade of brown, in the future it is going to mean a great deal more to all of us in many different ways. Solar energy is here to stay!

APPENDIX

This book is concerned with the facts about solar energy and its potential value in everyday applications. It will be appreciated by the reader that undertaking the installation of solar collectors can be both time-consuming and expensive. Although it is a task which can and has been undertaken by do-it-yourself exponents, the vast majority of flat-plate collectors are installed by specialist companies. The quality of workmanship that can be obtained from the professional companies has greatly improved in recent years, as a result of various measures taken by the Solar Energy Association. In addition, the preparation of a British Standard for solar collectors has made more people aware of the necessity to establish new criteria in this expanding home-improvement sphere of activity.

It should be borne in mind that most people who are considering using solar energy do so with the intention of saving on their energy bills. For this reason they are contemplating spending fairly substantial sums of money which they will not recoup for several years. Before such expenditure can be justified, every effort should be made to reduce energy wastage by other methods if they are cheaper and more cost-effective. It is, in fact, very easy to save much more than 10 per cent of the annual energy bill for very small cost and to recover even that in a very short period of time. And this applies to both domestic and industrial buildings.

In an uninsulated house 25 per cent of the heat escapes through the roof, yet this is about the cheapest place to insulate, 7.5-10 cm (3-4 inches) of glass fibre producing a saving of about 10 per cent in the total house energy bill. Another way to cut heat losses is to ensure that all pipes carrying hot water to the taps in bathrooms and kitchens are also insulated. The cold section of such pipes is known as the 'dead leg', and a fairly good idea of how long the 'dead legs' are can be obtained just by running the hot taps after

104

they have not been used for an hour and seeing how long it takes for hot water to come through. The cold water that is initially run out is former hot water that has lost its heat through the pipe. About 15 per cent of other heat losses can be eliminated by blocking off draughts around doors and windows using simple self-adhesive draught-excluding strips. Similarly it is worth considering the possibility of cavity wall insulation, as up to 35 per cent of the heat-loss from a house is through the walls. From a combination of these figures it can be seen that something of the order of 40-50 per cent of the total domestic energy bill can be saved by proper insulation alone. This gives a much more rapid return for the money spent and should always be carried out before spending money on solar radiation or double glazing.

Many of these points also apply to the working environment. Employers should not only be introducing insulation wherever possible, but also be carrying out energy consumption assessments on all their equipment and on each room or building. Factory campaigns on energy saving should be carried out with as much sense of purpose as are those for factory safety.

Any householder contemplating the installation of solar collectors and seeking further advice on the quality of products available should contact the Solar Trade Association Ltd., c/o The Building Centre, 26 Store Street, London, WC1E 7BT. This body will not only provide a list of its members, which includes details of the services they provide, but also runs a conciliation service between members and customers. It is working to raise the standards within the solar energy industry and to improve public knowledge about the subject, and has now introduced a strict code of practice for all its members.

Those readers interested in the wider technical and international aspects of solar energy should contact the International Solar Energy Society. The headquarters of this society is in Australia, but the United Kingdom Section (UK-ISES) is at 19 Albermarle Street, London W1X 3HA. Information about do-it-yourself systems of solar water heaters, solar roofs and other energy conservation techniques can be obtained from the National Centre for Alternative Technology, Llwyngwem Quarry, Machynlleth, Powys, Wales.

Additional information on the commercial and technical aspects of solar energy applications can be obtained from Dr. C.M.S. Johansson, Solar Energy Unit, Department of Mechanical Engineering and Energy Studies, University College, Newport Road, Cardiff CF2 1TA. This particular unit has been established

with the support of Government funds through the Department of
Industry to assist and encourage British industry and professional
engineers, scientists and designers to become actively involved in
solar energy applications and development. The Unit also publishes
a quarterly journal *Helios* dealing with all aspects of solar energy
research and its uses throughout the world.